AF407090

Sustainable Asset Management: AI & Blockchain Unleashed

Comprehensive Guide to Implementing Maintenance Strategies with AI, Blockchain, and Sustainability: Enhancing Efficiency and Reducing Downtime

Second Edition

Author: Prashant Sinha

Founder

algorithmicbias.ai

https://algorithmicbias.ai/

Email: ps@algorithmicbias.ai

About the Author(s)

Prashant Sinha, a passionate technophile and poker aficionado, stands at the forefront of AI innovation as the pioneering founder of Algorithmic Bias, a firm devoted to creating bespoke AI solutions across various sectors. His career spans a rich tapestry of roles and experiences, from hands-on software engineering to leadership positions in start-ups, culminating in establishing a thriving tech enterprise.

Prashant's distinct aptitude for technology and leadership has been evident throughout his professional journey. He has navigated the complex worlds of investment banking, e-commerce, cloud/SaaS services, and the public sector, always leaving an indelible mark of innovation and outstanding performance. In each role, Prashant has demonstrated his unique ability to foster innovation, facilitate effective teamwork, and deliver exceptional results.

One of the most telling aspects of Prashant's character is his long-term commitment to public service. Prashant Sinha has spent more than a decade working in asset management for a public wastewater utility, a testament to his dedication to contributing to the public good.

Prashant Sinha's depth of experience across diverse industries and his steadfast commitment to pushing the boundaries of technology make him uniquely qualified to guide readers through the intricacies of AI, Blockchain, and asset management in this insightful book. His journey is a testament to the power of innovation, collaboration, and public service – a power that he hopes to share with readers and inspire them to apply in their lives and careers.

Table of Contents

Foundations of Asset Management

Asset Management is a systematic process of cost-effectively operating, maintaining and upgrading assets. It is an interdisciplinary field that touches upon finance, business, and engineering principles. Asset management aims to ensure the optimal use of resources and maximize the value of assets while managing risk and providing the necessary level of service cost-effectively.

The concept of asset management has been around for centuries. It originated from the need to manage physical assets like buildings, machinery, and equipment. However, the concept has evolved and expanded over time to include intangible assets like patents, trademarks, and brand names.

Today, Asset Management is an integral part of any organization. For example, it is critical in the manufacturing, utilities, and transportation industries, where physical assets are crucial in service delivery. But it's equally relevant in sectors like IT and finance, where intangible assets dominate.

At its core, Asset Management involves balancing costs, opportunities, and risks against the desired performance of assets to achieve an organization's objectives. Asset Management includes managing the entire lifecycle of assets, from the design and acquisition stage to operation, maintenance, and eventual disposal.

In the current scenario, Asset Management is not just about maintaining and tracking physical assets; it's about strategic decision-making and long-term planning. The focus is on understanding how assets contribute to an organization's performance and how to manage them to improve it.

Effective Asset Management requires a deep understanding of how different assets contribute to an organization's goals and the interdependencies between them. It involves making informed decisions about where to invest in new assets, how to maintain existing assets, when to refurbish or replace assets, and when to retire them.

Technological advances have greatly enhanced our ability to manage assets in recent years. Technologies like the Internet of Things (IoT), Artificial Intelligence (AI), and blockchain are revolutionizing Asset Management. For example, IoT sensors allow for real-time monitoring of assets, AI enables predictive maintenance, and blockchain provides secure, transparent tracking of asset lifecycles.

However, as with any discipline, Asset Management also has its challenges. These include ensuring compliance with regulations and standards, managing the impact of external factors like climate change, and dealing with data quality and integration issues.

Despite these challenges, the importance of Asset Management continues to grow. As we move towards a more resource-constrained world, the ability to manage and optimize the use of assets is becoming increasingly critical. And as technology continues to evolve, the future of Asset Management looks promising, with even more excellent opportunities for efficiency and innovation on the horizon.

Types of Assets

Physical Assets: These are tangible items such as buildings, machinery, equipment, and vehicles. Managing these assets involves tasks like maintenance, inspection, and replacement.

Financial Assets: These include things like cash, investments, accounts receivable, and other financial resources. Financial asset management involves strategies for maximizing returns and minimizing risk.

Digital Assets: These are electronic files or data, such as software, databases, and digital media. Digital asset management involves organizing, storing, and retrieving these assets and managing access rights and permissions.

Human Assets: These are a company's employees' skills, talents, and abilities. Human asset management, often referred to as human resources management, involves recruiting, training, and retaining employees.

This book's scope will focus primarily on physical assets and their management and maintenance. While all forms of assets, including financial, intangible, and human resources, play a critical role in any organization's success, this book will concentrate on the principles and practices of managing physical assets.

Physical assets include machinery, buildings, infrastructure, and other tangible properties owned and used by an organization to generate revenue or provide services. These assets require careful management to ensure optimal performance, minimize downtime, and extend their useful life. In the context of climate action, physical asset management gains even more significance as these assets are often significant contributors to greenhouse gas emissions and energy consumption.
Physical asset management and maintenance are key to enhancing operational efficiency and driving sustainability. This dual focus aligns organizations with their environmental goals and responsibilities while ensuring financial viability.

By limiting the scope to physical asset management, we can delve deeper into the complexities of this field, discussing strategies, challenges, and opportunities in a focused, detailed manner. This approach allows us to explore quantitative methods, climate data application, and how these intersect with the critical function of asset management.

The following chapters will guide you through the fascinating landscape of physical asset management, where scientific data, innovative management strategies, and climate action intertwine. By the end of this book, you will be equipped with the knowledge and tools necessary to manage physical assets in a climate-conscious world effectively.

The Birth of Asset Management

Asset management has been around for as long as there have been assets. In ancient times, these assets were simple: they were primarily land, livestock, and tools. However, managing these assets effectively was crucial to survival and prosperity. The earliest forms of asset management were born out of necessity, as people needed to keep track of what they owned, ensure their assets were being used efficiently, and make decisions about when to repair or replace them.

As societies evolved, so did the nature of assets. With the advent of trade, currency, and commodities became significant assets, leading to the development of financial asset management. Financial asset management involved managing wealth by investing in assets that could grow in value, such as gold, land, or stocks in a trading enterprise.

In the industrial era, businesses began recognizing the importance of managing physical assets such as machinery and buildings. In addition, the industrial era was a time of rapid technological advancement, and businesses that could effectively manage their assets had a competitive advantage. This led to the birth of corporate asset management, with businesses developing systems and processes to track, maintain, and replace their assets.

These early forms of asset management were primarily reactive and focused on managing the immediate needs of assets. However, over time, people realized the benefits of a more proactive approach. This led to developing strategies for maintaining assets in optimal condition, extending their life, and improving their performance. This proactive approach to asset management is the foundation of modern asset management practices.

Asset Management in the Industrial Age

The Industrial Revolution marked a significant turning point in history. The widespread adoption of new manufacturing processes revolutionized many industries, leading to unprecedented growth in the production of goods. This rapid industrialization brought with it the need to effectively manage a new type of asset: machinery.

In the Industrial Age, machinery became the backbone of production. The replacement of manual labor with machines in many industries filled factories and resulted in a surge in productivity. However, these machines, often complex and expensive, required careful management to keep them running smoothly and to protect the significant investment they represented.

In this era, asset management evolved from merely tracking what was owned to strategically managing assets to maximize their value. For example, asset management meant maintaining machines to prevent breakdowns and optimizing their usage to improve productivity. In addition, businesses began to understand the importance of lifecycle management – the practice of considering the entire lifespan of an asset from procurement to disposal when making decisions about its management.

One of the significant developments during this era was the introduction of planned maintenance schedules. Instead of waiting for a machine to break down before repairing it, businesses started to perform regular, scheduled maintenance to prevent breakdowns from occurring in the first place. This proactive approach to asset management represented a significant shift in thinking and laid the foundation for many modern asset management practices.

The Industrial Age also saw the rise of specialization in asset management roles. Specific individuals or teams were assigned the responsibility of maintaining assets, leading to the emergence of new professions such as maintenance engineering.
The advent of the Industrial Age, with its focus on machinery and manufacturing, transformed asset management into a critical business function. This trend has only intensified as we've moved into the modern era.

The methods of managing physical assets have evolved significantly over time, primarily driven by technological advancements and shifts in business practices and economic conditions. This chapter traces the journey of physical asset management from its most basic forms to the complex systems in use today, setting the stage for an in-depth discussion of the role of artificial intelligence in modern asset management.

Early Asset Management

In the industry's early days, asset management was a largely manual process. Assets were tracked using paper-based systems, and maintenance was typically reactive, carried out only when something broke down. Unfortunately, this approach was labor-intensive and prone to error, leading to inefficiencies, unexpected equipment failures, and costly downtime.

The Advent of Computerized Systems

The development of computer technology in the 20th century revolutionized asset management. Computerized Maintenance Management Systems (CMMS) began to emerge, enabling organizations to track their assets digitally, schedule preventive maintenance, manage work orders, and more. These systems significantly improved efficiency and reduced the risk of unexpected equipment failures.

The Shift to Proactive Maintenance

Over time, the asset management philosophy shifted from a reactive approach to a more proactive one. For example, instead of waiting for equipment to break down, organizations started implementing preventive maintenance strategies, scheduling regular inspections and maintenance to keep equipment in good working order and prevent failures before they occur.

The Rise of Predictive Maintenance

Predictive maintenance has become increasingly popular with advancements in sensor technology and data analysis. This approach uses data from sensors installed on equipment to predict when a failure is likely to occur, allowing maintenance to be scheduled just in time to prevent the failure. As a result, predictive maintenance further improves efficiency and reduces downtime.

Conclusion: The Advent of AI in Asset Management

Today, we stand on the cusp of a new revolution in asset management with the advent of artificial intelligence. AI has the potential to take predictive maintenance to a new level, learning from past data to make increasingly accurate predictions about future failures. It also offers the possibility of cognitive maintenance, where AI systems learn and reason like human experts, making complex maintenance decisions and even automatically adjusting operating conditions to optimize equipment performance.

In the following chapters, we will delve deeper into these exciting new developments, exploring the potential of AI in asset management, the challenges it presents, and strategies for successfully implementing AI in your organization.

I cannot overstate the impact of technological advancements on asset management. The move from manual to automated processes has transformed how organizations track, manage, and optimize their assets. As technology has evolved, so too has the complexity and capability of asset management systems.

The advent of computer systems in the mid-twentieth century marked a significant leap forward for asset management. Computers allowed for digitizing and centralizing asset data, making it more accessible and easier to analyze. This, in turn, facilitated more informed decision-making around asset usage and maintenance.

Another significant technological development was the advent of the Internet. The ability to connect and share data across computer networks has had profound implications for asset management. For example, it enabled real-time tracking and monitoring of assets, regardless of physical location. This development led to the rise of Computerized Maintenance Management Systems (CMMS) and, later, more comprehensive Enterprise Asset Management (EAM) systems. These systems provide a centralized platform for managing all aspects of an organization's assets, from tracking and maintenance to lifecycle management and cost analysis.

The evolution of sensor technology and the emergence of the Internet of Things (IoT) have taken asset management to new heights. Sensors attached to assets can now collect vast amounts of real-time data about their condition and performance. When analyzed using advanced analytics and machine learning algorithms, this data can provide valuable insights, enabling predictive maintenance practices. In predictive maintenance, we identify and resolve potential issues before they can cause a breakdown, significantly reducing downtime and maintenance costs.

The rise of mobile technology has also had a significant impact, enabling asset managers to access and update asset information on the go. For example, mobile apps can now allow maintenance personnel to instantly report issues, access maintenance manuals, and update work order statuses from the field.

Technological advancements have improved the efficiency and effectiveness of asset management practices and opened up new possibilities. With the advent of technologies such as blockchain and artificial intelligence, the future of asset management promises to be even more dynamic and transformative.

As we moved into the 21st century, the rise of modern asset management took hold, characterized by a greater focus on strategic, holistic, and data-driven practices. Rapid technological advancements largely facilitated this shift, enabling the collection, analysis, and utilization of vast data.

Modern asset management is not just about maintaining and repairing equipment; it's about managing assets throughout their lifecycle, from acquisition and deployment to maintenance and decommissioning. It's about optimizing the performance and value of assets while minimizing the total cost of ownership.

Enterprise Asset Management (EAM) systems have become a staple in modern asset management. These comprehensive platforms provide a centralized repository for all asset-related data, facilitating effective tracking, maintenance, and lifecycle management. In addition, EAM systems offer features such as work order management, predictive maintenance, asset tracking, inventory management, and more.

Integrating the Internet of Things (IoT) into asset management has revolutionized the field. IoT devices and sensors actively collect real-time data about asset performance and conditions, enabling the anticipation and prevention of equipment failures. This approach, known as predictive maintenance, has become a significant trend in modern asset management.

Artificial intelligence (AI) and machine learning have also started to play an increasingly important role. These technologies enable the analysis of vast amounts of asset data to uncover patterns and insights that were previously out of reach. As a result, they can predict asset failures, optimize maintenance schedules, and even automate specific decision-making processes.

Another notable trend is the growing emphasis on sustainability and green asset management. This is in response to increasing environmental concerns and regulatory pressures. Modern asset management often includes strategies for reducing energy usage, minimizing waste, and extending asset lifecycles.

Lastly, mobile technology and cloud computing have made asset management more flexible and accessible. Asset managers can now access asset data and perform critical tasks from anywhere, at any time. Cloud-based asset management solutions offer scalability, flexibility, and cost savings, making them increasingly popular.

In summary, the ascent of contemporary asset management has witnessed a transition from reactive to proactive and predictive strategies propelled by the integration of technology and data. This domain remains in a state of constant evolution, embracing new technologies and methodologies as they emerge.

Several emerging trends and technologies shape the future of asset management, and understanding these trends is crucial for any organization that wants to stay ahead of the curve.

Trend 1: Internet of Things (IoT)

The Internet of Things (IoT) refers to the network of physical devices connected to the Internet, capable of collecting and sharing data. With IoT, assets can "talk" to us and each other, providing real-time data on their condition and performance. The data can be utilized to predict failures, optimize maintenance schedules, and enhance asset utilization.

Trend 2: Artificial Intelligence (AI)

As we've touched upon in the previous chapters, AI plays a critical role in the future of asset management. AI algorithms can analyze large amounts of data from IoT devices to predict asset behavior, optimize maintenance schedules, and make data-driven decisions. Additionally, AI can assist in identifying and mitigating algorithmic bias, ensuring fairness and accuracy in decision-making.

Trend 3: Augmented Reality (AR) and Virtual Reality (VR)

AR and VR technologies are starting to make their way into asset management. They can be used for training purposes, to simulate maintenance tasks, and to visualize asset data intuitively. For example, a technician wearing AR glasses could see a digital overlay of an asset's maintenance history and current condition while inspecting.

Trend 4: Blockchain

While still in the early stages of adoption, blockchain technology has potential applications in asset management. Blockchain could provide a secure and transparent way to track asset transactions and changes in ownership. It could also automate contract execution and payment through smart contracts.

Trend 5: Sustainability and ESG

Environmental, social, and governance (ESG) considerations are increasingly important in asset management. As a result, organizations are being held accountable for the environmental impact of their assets and operations, and there is a growing demand for sustainable and socially responsible asset management practices.

In the next part of the book, we'll delve deeper into how AI is revolutionizing asset management. We'll explore how AI-driven predictive, condition-based, and cognitive maintenance transform how assets are managed, leading to improved efficiency, cost savings, and better decision-making. I will also shed light on how AI systems are not free from bias and how ensuring fairness in AI-driven asset management is essential. The intersection of AI and asset management is fascinating, teeming with opportunities and challenges. Stay tuned as we delve into these aspects in the subsequent chapters.

Physical assets play a critical role in the operation and success of organizations across different industries. Physical assets are integral components of their respective value chains, from manufacturing and healthcare to transportation and utilities. Here is a look at the role of physical assets in different industries:

Manufacturing Industry: The manufacturing industry heavily relies on physical assets, such as machinery and equipment, to produce goods. Assets like assembly lines, robotic arms, conveyor belts, and CNC machines are the lifeblood of manufacturing operations. Their performance directly impacts productivity, quality, and safety in the workplace.

Healthcare Industry: In the healthcare industry, physical assets range from medical equipment like MRI machines, CT scanners, and surgical instruments to the IT infrastructure that supports electronic health records and telemedicine. These assets are vital for delivering quality patient care and ensuring efficient operations.

Transportation Industry: In the transportation industry, physical assets such as vehicles, aircraft, ships, and infrastructure (roads, bridges, airports, and seaports) are essential for moving people and goods. The condition and availability of these assets can significantly impact service reliability, safety, and customer satisfaction.

Utility Industry: Utilities like electricity, water, and gas depend on a vast network of physical assets, including power plants, pipelines, transformers, and water treatment facilities. Effective asset management in this industry is crucial for maintaining reliable service, complying with regulations, and managing environmental impact.

Oil and Gas Industry: In the oil and gas industry, physical assets include drilling rigs, pipelines, refineries, and storage tanks. These assets are critical for exploration, extraction, refining, and distribution. Consequently, their performance and reliability can significantly affect operational efficiency, safety, and compliance.

Real Estate and Facilities Management: In this sector, physical assets primarily consist of buildings and facilities. The maintenance and management of these assets, including HVAC systems, elevators, and safety systems, are crucial for occupant comfort, safety, and compliance with building codes.

In each of these industries, managing physical assets effectively is paramount. It requires a strategic approach considering the entire asset lifecycle, from acquisition and deployment to maintenance and eventual decommissioning. In addition, technological advances, such as IoT,

AI, and predictive analytics, enable more efficient, data-driven asset management practices that can improve asset performance, extend asset life, and reduce costs.

Physical assets are the backbone of production processes in the manufacturing industry, essential in transforming raw materials into finished goods. They range from large machinery and equipment used in production lines to smaller tools and devices for quality assurance and control. Here's a closer look at the role of physical assets in the manufacturing industry:

Production Machinery and Equipment: These are the main drivers of any manufacturing process. Depending on the nature of the industry, these could include milling machines, lathes, welding machines, assembly lines, conveyor belts, robotic arms, and other specialized machinery. These assets are responsible for various stages of production, such as cutting, shaping, assembling, and finishing of products.

Quality Assurance and Control Instruments: These assets are used to ensure the products meet the required quality standards. They may include precision measurement tools, spectrometers, scanners, and various types of testing equipment. Regular use and maintenance of these tools are crucial for maintaining product quality and customer satisfaction.

Supporting Infrastructure: Physical assets in manufacturing also include the facilities and infrastructure that support production, such as warehouses, factories, power systems, and HVAC systems. These assets provide the environment for manufacturing processes to occur efficiently and safely.

Information Technology (IT) Equipment: In the age of Industry 4.0, IT equipment forms a critical part of the physical asset base in manufacturing. This includes servers, computers, networking devices, and factory automation systems. These assets facilitate the digitization of manufacturing processes, enabling more efficient production, real-time monitoring, and advanced analytics.

Vehicles: Many manufacturing firms also possess a fleet of vehicles for transporting raw materials, finished goods, or even employees. These assets need to be effectively managed to ensure timely deliveries and efficient logistics.

In all cases, these assets require regular maintenance and periodic upgrades to keep up with evolving industry standards and technological advancements. As a result, asset management plays a central role in maximizing asset utilization, minimizing downtime, ensuring safety, and achieving cost-effectiveness in manufacturing. In addition, with the advent of technologies like the Internet of Things (IoT), Artificial Intelligence (AI), and predictive analytics, manufacturers are now better equipped to monitor asset performance in real-time, predict potential failures, and optimize maintenance schedules. Asset management improves asset reliability, longer life, and overall operational efficiency.

Public utilities and services, encompassing sectors such as water, electricity, gas, telecommunications, and public transport, rely heavily on their physical assets to provide essential services to the public. These assets' nature, role, and importance vary across different utilities. Let's delve into the part of physical assets in these sectors:

Water and Sewerage Systems: In the water sector, physical assets include reservoirs, treatment plants, pumping stations, pipelines, and valves, among others. These assets are responsible for the collection, treatment, and distribution of water, as well as the collection and treatment of wastewater. Therefore, regular maintenance and upgrades are crucial to prevent leaks, contamination, and service disruptions.

Electricity and Gas Infrastructure: In the energy sector, physical assets include power plants, substations, transformers, transmission and distribution lines, and gas pipelines. These assets ensure the generation, transmission, and distribution of electricity and gas to consumers. Therefore, these infrastructures require vigilant asset management to provide a reliable and uninterrupted energy supply.

Telecommunications Infrastructure: Assets in the telecommunications sector include data centers, servers, network infrastructure such as cell towers and fiber optic cables, and customer premises equipment like routers and modems. These assets enable the transmission of voice, data, and video services, which is critical in connecting people and businesses.

Public Transport Systems: In public transport, assets include vehicles (buses, trams, trains, ferries), stations, terminals, tracks, signaling systems, and maintenance facilities. These assets are essential for providing public transportation services and must be properly managed to ensure safety, reliability, and efficiency.

Managing these assets is particularly crucial because of the vital public services they support. Asset failure can lead to service disruptions, safety incidents, and other consequences affecting many people. Hence, effective asset management strategies, including preventive and predictive maintenance, risk management, and asset lifecycle management, are essential in the public utilities and services sector.

The advent of advanced technologies like IoT, AI, and blockchain has greatly enhanced asset management in public utilities and services. They enable real-time asset monitoring, predictive maintenance, improved decision-making, enhanced transparency and security, driving efficiency, reliability, and customer satisfaction.

Physical Assets in Wastewater Treatment Facilities

In wastewater treatment facilities, physical assets play a crucial role in protecting public health and the environment by treating and purifying used water before it's returned to the environment or reused. These facilities consist of interconnected physical assets, each performing a specific role in the treatment process. Let's explore the different types of physical assets found in wastewater treatment facilities and their functions:

Preliminary Treatment Assets: These include screens, grit chambers, and grease traps. Screens filter out large debris, while grit chambers allow sand and other small particles to settle. Grease traps prevent fats, oils, and grease from entering the treatment system. These assets are essential to prevent damage to downstream equipment and processes.

Primary Treatment Assets: Primary clarifiers or sedimentation tanks are used in this stage. They allow the organic solid matter to settle to the bottom, forming sludge, while the oil and grease float to the top as scum. Regularly maintaining these tanks is crucial to prevent buildup and ensure efficient operation.

Secondary Treatment Assets: This stage involves biological processes to remove dissolved and suspended organic matter. It usually involves assets like aeration tanks and secondary clarifiers. Aeration tanks use microorganisms to consume the organic matter, and secondary clarifiers allow these microorganisms to settle as sludge. Maintaining aeration equipment is essential to ensure sufficient oxygen supply for the microorganisms.

Tertiary Treatment Assets: Tertiary treatment assets may be required depending on the intended use of the treated water, including sand filters, activated carbon filters, disinfection equipment (such as UV light systems or chlorination equipment), and membranes for advanced filtration or reverse osmosis. These assets provide additional treatment to meet specific water quality standards.

Sludge Treatment Assets: These include sludge thickeners, digesters, dewatering equipment, and incinerators or biosolids handling facilities. These assets treat the sludge produced during primary and secondary treatment, reducing its volume and making it safer for disposal or beneficial use.

Supporting Assets: These include pumps, blowers, motors, pipes, control systems, and SCADA systems. They support the operation of the primary treatment assets and require regular inspection, testing, and maintenance to ensure the smooth operation of the facility.

Effective asset management is crucial in wastewater treatment facilities due to their vital role in public health and environmental protection. This involves regular preventive maintenance, condition monitoring, lifecycle management, and risk management. Advanced technologies like AI, IoT, and blockchain can significantly enhance these efforts by enabling real-time monitoring, predictive maintenance, improved decision-making, and enhanced transparency and security.

Physical Assets in the Automobile Industry

Physical assets play a pivotal role in the automobile industry, helping to bring vehicles from conceptual design to the roads. These assets span multiple domains, from manufacturing facilities and assembly lines to diagnostic equipment and even vehicles. Let's delve into the essential physical assets in the automobile industry:

Manufacturing Equipment: This includes machinery and tools used in the production of vehicles, such as presses, lathes, milling machines, welding robots, and assembly lines. These assets are integral to transforming raw materials into finished products and need regular maintenance to ensure operational efficiency and quality control.

Assembly Lines: Assembly lines are a critical physical asset in the automobile industry. These assets, comprising various automated and semi-automated machines, streamline the production process by moving a product along a set path where different components are added until the final product is assembled. The advent of robotics and automation has transformed assembly lines, making them more efficient and reducing human error.

Testing and Diagnostic Equipment: Quality control in the automobile industry is crucial; therefore, testing and diagnostic equipment are valuable assets. These include dynamometers, emission testers, crash testing equipment, and various other diagnostic tools used to ensure vehicle safety, performance, and reliability.

Storage and Distribution Facilities: Warehouses, distribution centers, and logistics vehicles are vital in storing and transporting vehicles from the factory to dealerships. Efficiently managing these assets can significantly impact delivery times and customer satisfaction.

Automobiles Themselves: The vehicles produced are, of course, significant physical assets in the automobile industry. These assets represent substantial financial value from the moment they roll off the production line until they're sold. Therefore, they must be effectively managed to prevent damage, theft, or depreciation.

Facilities and Real Estate: Factories, offices, research & development centers, and dealerships all form part of the physical assets that need effective management. These assets should provide employees with a safe, efficient, and productive environment while aligning with the company's sustainability and environmental goals.

Information Technology Assets: In the era of Industry 4.0, IT assets have become critical in the automobile industry. These include computer systems, servers, software, and data centers that support design, production, inventory management, sales, and after-sales services. Also, with the rise of connected cars, IT assets are increasingly embedded within the vehicles themselves, necessitating new approaches to asset management.

Asset management in the automobile industry is crucial to ensure efficiency, quality, and cost-effectiveness. It involves regular maintenance, timely upgrades, lifecycle analysis, and strategic decommissioning of assets. Today, technologies like IoT, AI, and blockchain are harnessed to optimize asset management, offering real-time monitoring, predictive maintenance, and enhanced supply chain transparency.

Harnessing Inventory Management for Enhanced Asset Management

Introduction

Inventory management is a critical operational process for organizations of all sizes and industries. It is the systematic approach to sourcing, storing, and selling inventory, from raw materials to finished goods. Inventory management aims to strike a balance between managing costs and meeting customer demands.

Effective inventory management ensures enough stock to meet customer needs while minimizing the cost of holding, ordering, and shipping inventory. It encompasses a range of tasks, such as:

- **Forecasting demand:** Predicting how much inventory will be needed in the future.

- **Determining when to reorder products:** Deciding when to order new inventory so there is always enough to meet demand.

- **Keeping track of inventory levels:** Maintaining accurate records of how much inventory is on hand and where it is located.

- **Analyzing sales patterns:** Tracking sales data to identify trends and patterns that can be used to improve inventory management decisions.

Inventory management is a complex and challenging process, but it is essential for the success of any organization that sells products or services. Organizations can improve their bottom line, reduce risk, and improve customer satisfaction by effectively managing inventory.

Inventory management is a fundamental aspect of efficient asset management and maintenance. It involves the planning, organizing, and controlling inventory items, which in the context of asset management, refer to essential components such as spare parts, replacement units, and maintenance tools. The main objective is to ensure the correct quantity of these items is readily available at the right time and place to avoid unnecessary downtime, extend asset lifespan, and maintain operational efficiency.

Effective inventory management is crucial in ensuring that the required parts are available when needed to repair or replace components of assets such as machinery or equipment. This helps to reduce the time taken to repair or replace components, resulting in minimal disruption to operations. Proper inventory management can also help lower the carrying costs associated with holding excess inventory and prevent stock-outs that might delay maintenance work.

For organizations, intertwining inventory management and asset maintenance forms a symbiotic relationship. Inventory stocked with necessary parts and tools directly fuels the asset maintenance process. Conversely, an effective maintenance strategy can inform the inventory needs, reducing instances of overstocking or understocking.

In the dynamic and fast-paced environment of asset management and maintenance, there is a growing recognition of the need for intelligent systems to manage inventory effectively and efficiently. This is where cutting-edge technologies such as artificial intelligence (AI) and blockchain emerge, enhancing traditional methods with predictive capabilities and secure, decentralized control.

This chapter provides a comprehensive overview of the importance of inventory management in the context of asset management and maintenance. It will delve into the role and significance of efficient inventory management in ensuring optimal asset performance, the challenges faced, and how emerging technologies can solve them.

Inventory management and **asset management** are two closely related disciplines essential for a business's success. Both disciplines are concerned with the flow of goods but focus on different aspects of the process.

Inventory management is the process of ordering, storing, using, and selling a company's inventory. It is typically focused on short-term goods that are used in the production or sale of products.

Asset management is acquiring, maintaining, and disposing of a company's assets. It is typically focused on long-term goods used in a business's operation.

Despite their different focuses, inventory management and asset management are interconnected. For example, a company that manufactures products needs to ensure that it has the necessary parts and materials in stock for routine maintenance, unexpected repairs, and replacements. This is where inventory management comes in.

Here are some ways in which inventory management and asset management are related:
- **Optimal operation and maintenance of assets:** Effective inventory management ensures that the necessary parts and components are available when needed for the routine maintenance or repair of assets, minimizing downtime and maintaining operational efficiency.
- **Cost management:** Both inventory and assets represent significant investments for organizations. Effective management of both helps optimize costs — inventory management minimizes the costs associated with holding too much or too little inventory. In contrast, asset management maximizes the value derived from assets throughout their lifecycle.
- **Risk management:** Both practices are crucial for risk management. Without effective inventory management, organizations risk operational delays due to the unavailability of necessary parts. Similarly, assets may break down or fail without proper asset management, disrupting operations.
- **Data-driven decisions:** Both asset and inventory management can significantly benefit from data analytics. Accurate, real-time data can inform better decision-making, such as predicting when an asset may fail and require maintenance or determining optimal reorder levels for inventory.

In short, inventory management and asset management are two disciplines that are closely related and essential for the success of a business. By effectively managing their inventory and assets, businesses can improve their bottom line, reduce risk, and improve customer satisfaction.

Inventory management plays a fundamental role in **asset maintenance** operations in several capacities:
- **Supporting scheduled maintenance:** Efficient inventory management ensures that the necessary parts and materials are available when scheduled maintenance activities are due. This includes routine servicing, component replacements, and other regular tasks, ensuring minimal disruption to operations.
- **Enabling unscheduled maintenance:** In the case of unexpected equipment failure or malfunction, a well-managed inventory can provide quick access to necessary parts, minimizing downtime and potential productivity loss.
- **Facilitating predictive maintenance:** By integrating inventory management with predictive maintenance systems, organizations can anticipate the need for specific components or consumables and manage inventory accordingly. This allows organizations to purchase parts precisely when needed, reducing inventory holding costs.
- **Optimizing costs:** Good inventory management can significantly reduce costs associated with overstocking or understocking. Overstocking can lead to wasted capital and storage costs, while understocking can result in expedited shipping costs or production downtime.
- **Contributing to compliance:** In certain industries, regulatory bodies may require firms to maintain specific components or materials in inventory. Effective inventory management ensures compliance with these regulations, avoiding potential fines or sanctions.
- **Enhancing operational efficiency:** By ensuring that parts and materials are available when needed, inventory management reduces the risk of work stoppages, helps maintain equipment in good working condition, and ultimately enhances operational efficiency.
- **Supporting decision-making:** A well-structured inventory management system can provide valuable data for decision-making processes, helping to forecast demand, evaluate supplier performance, identify cost-saving opportunities, and plan for future capacity needs.

In essence, inventory management acts as a significant enabler of efficient and effective asset maintenance, directly contributing to operational reliability and cost optimization.

Ensuring the availability of spare parts is a crucial role of inventory management in asset maintenance. In preventive and corrective maintenance activities, having the right parts at the right time is crucial. This function involves several specific tasks:

- **Forecasting needs:** Inventory management systems must forecast the need for spare parts based on the historical usage of spare parts and the maintenance schedule of assets. This projection enables the system to maintain a balanced stock that avoids overstocking (which can lead to increased carrying costs and obsolescence) and stockouts (which can delay maintenance activities).
- **Managing suppliers:** Inventory management involves building and maintaining relationships with reliable suppliers who can provide quality parts at reasonable prices within a short lead time. It also involves evaluating supplier performance regularly and promptly addressing any issues to avoid supply disruptions.
- **Optimizing inventory levels:** This involves maintaining a safety stock to handle fluctuations in demand or disruptions in supply. The safety stock level depends on factors like the criticality of the part, lead time, and the cost of stockouts. A consignment inventory agreement with suppliers may be appropriate for expensive parts that are rarely used.
- **Organizing storage:** Effective inventory management also involves organizing the storage of spare parts in a way that makes retrieval efficient and prevents damage or loss. This includes implementing effective warehouse management practices like barcoding, implementing FIFO (First-In, First-Out) where applicable, and regularly auditing physical inventory.
- **Coordinating with maintenance planning:** Close coordination with the maintenance planning function is necessary to align spare part availability with maintenance schedules. This can involve adjusting inventory levels based on planned maintenance activities or coordinating with suppliers to expedite deliveries when unexpected maintenance needs arise.

Inventory management plays a vital role in minimizing equipment downtime, improving operational efficiency, and ultimately impacting an organization's bottom line by ensuring the availability of spare parts.

Reducing downtime is a crucial aspect of the role of inventory management in asset maintenance. Downtime in an industrial context refers to periods when an asset is unavailable or non-operational due to unforeseen breakdowns, scheduled maintenance, or equipment failures. Extended periods of downtime can lead to significant operational losses and decreased productivity, directly impacting a company's bottom line.

Here are some ways that efficient inventory management can contribute to reducing downtime:

- **Availability of critical spare parts:** Inventory management systems can significantly reduce the time taken for repair activities by ensuring that necessary spare parts are available when needed. This is particularly important when specialized parts are needed, which may have extended lead times if not stocked in advance.

- **Predictive inventory management:** Using historical data and predictive analysis, inventory management can forecast when specific parts might be needed based on previous maintenance activities and the operational lifespan of various components. By preparing for these eventualities in advance, you can minimize the time a machine is out of action.
- **Effective coordination:** Inventory management involves coordinating with other functions like procurement, logistics, and maintenance planning. Ensuring smooth communication and collaboration between these can speed up getting an asset back online.
- **Consignment inventory:** Consignment inventory agreements can be put in place with suppliers for highly expensive or rarely used parts. Under this agreement, the supplier stocks items at the company's location, but the company only pays for the parts when they are used. This approach can ensure the availability of specialized or expensive parts without tying up capital.
- **Just-in-time (JIT) inventory management:** This approach closely monitors inventory levels and reorders stock just in time to meet maintenance needs. While this method requires highly accurate demand forecasting and reliable suppliers, it can significantly reduce inventory holding costs and ensure parts are available when needed.

In these ways, effective inventory management can significantly reduce downtime, thereby enhancing productivity and overall operational efficiency.

Optimizing Repair vs. Replace Decisions

Inventory management is crucial in helping maintenance teams make optimal repair vs. replace decisions. The decision to repair a failing part or to replace it entirely is often complex, balancing factors such as the cost of the part, the cost of labor, the impact of downtime, and the potential for a repaired part to fail again.

Inventory management can provide valuable data and insights to guide these decisions:

- **Availability of parts:** If the required parts for a repair are readily available in inventory and the repair can be done quickly and cost-effectively, it may lean the decision towards a repair. Conversely, if the part is not in stock and must be ordered with a long lead time, replacement of the whole unit might be a better option.
- **Cost information:** Inventory management systems often have detailed cost information about parts, which can be used in cost-benefit analysis. For instance, if a spare part costs nearly as much as a new unit, it would make sense to replace the entire unit rather than repair it.
- **Historical data:** Inventory management systems can provide data on the frequency of past repairs, which can help predict the likelihood of future failures. If a part frequently needs repairs, replacing the entire unit might be more economical in the long term.

- **Condition of spare parts:** The condition of parts in inventory could also influence the decision. For example, replacement could be a better option if parts have been in storage for a long time and might have deteriorated or become obsolete.
- **Vendor management:** Inventory management also involves maintaining relationships with vendors. Good vendor relationships can make it easier to return parts or negotiate discounts, which could sway a repair/replace decision.

By providing necessary data and enabling these analyses, inventory management is essential in optimizing repair vs. replace decisions, leading to more efficient and cost-effective maintenance operations.

Maintenance Strategies for Physical Assets

Maintenance is a critical component of successful asset management. Organizations can employ different strategies based on the nature of the asset, its importance to operations, and the cost-effectiveness of various maintenance approaches. Below are some of the most common strategies used for maintaining physical assets:

Reactive Maintenance (Run-to-Failure): This is the most straightforward maintenance strategy. In this approach, assets are allowed to operate until they break down. At this point, reactive maintenance is performed to repair the asset and restore it to operational condition. While this method can be cost-effective for non-critical assets or those with low repair costs, it can lead to higher downtime and unpredictable maintenance expenses.

Preventive Maintenance (Time-based or Usage-based): Under this strategy, maintenance personnel performs maintenance at regular intervals based on time (e.g., monthly, annually) or usage (e.g., after every 1,000 hours of operation). Preventive maintenance aims to catch and correct minor issues before they become major problems that can cause an asset to fail.

Condition-based Maintenance (CBM): This advanced maintenance strategy involves actively monitoring the actual condition of the asset to determine the appropriate timing for maintenance activities. CBM uses sensors and other data collection tools to identify signs of potential failure (e.g., unusual vibrations, temperature variations, abnormal noise levels). Maintenance is then scheduled based on these indicators, which can help avoid unnecessary maintenance and minimize downtime.

Predictive Maintenance (PdM): This strategy uses advanced analytics and machine learning algorithms to predict when an asset is likely to fail based on historical data and real-time condition monitoring. This allows for maintenance to be scheduled just before the predicted failure, maximizing the useful life of the asset and minimizing downtime.

Reliability Centered Maintenance (RCM): This comprehensive and systematic maintenance approach focuses on ensuring assets continue to do what their users require in their present operating context. It involves identifying the functions of an asset, how it can fail, the causes of each failure mode, and the consequences of each failure. Based on this analysis, the most appropriate maintenance strategy (reactive, preventive, CBM, or PdM) is chosen for each asset.

Cognitive Maintenance: This is the latest evolution in maintenance strategies, leveraging artificial intelligence (AI) and machine learning to optimize maintenance decisions. Cognitive maintenance systems can process vast amounts of data from various sources, learn from this data, and provide predictive insights and recommendations for maintenance.

Each of these strategies has its advantages and disadvantages. The choice of strategy will depend on various factors, including the type of asset, its operational importance, the cost of downtime, and the organization's overall asset management objectives. By selecting and implementing the right maintenance strategy, organizations can maximize the reliability and lifespan of their physical assets, reduce maintenance costs, and improve operational efficiency.

Understanding the true cost of an asset is vital for organizations, particularly when making capital expenditure (CAPEX) decisions. These costs extend beyond the initial acquisition price and encompass the entire asset lifecycle. Below are the key considerations when evaluating the true cost of an asset for CAPEX projects:

Acquisition Costs: This includes the price of purchasing the asset and any associated costs such as transportation, installation, and initial configuration or customization. For certain assets, factoring in licensing or permit fees may also be necessary.

Operating Costs: Operating costs encompass the regular expenses associated with using the asset, such as energy costs, consumables, and operator wages for physical assets. It may include subscription fees or data costs for software or digital assets.

Maintenance Costs: Regular maintenance is crucial to keep an asset operating at peak efficiency and to prolong its useful life. Maintenance costs include routine servicing, replacement parts, and potentially the labor cost of the maintenance team.

Downtime Costs: When an asset is not operational due to maintenance or unforeseen breakdowns, significant costs can be associated with the loss of production or service. It's essential to account for potential downtime when considering the true cost of an asset.

Depreciation and Amortization: Over time, most assets lose value through wear and tear, obsolescence, or depletion. This decrease in value, or depreciation, should be factored into the total cost of the asset.

Disposal Costs: At the end of its useful life, it is essential to safely and responsibly dispose of an asset. Depending on the asset's nature, there may be costs associated with dismantling, recycling, or disposing of it in a manner that complies with environmental regulations.

Opportunity Costs: This represents the cost of the next best alternative foregone. When investing capital in a specific asset, it is crucial to consider the missed opportunities.

Risk Costs: Every asset brings with it certain risks, such as the risk of failure, the risk of becoming obsolete, or even the risk of causing accidents or environmental damage. Factoring these potential costs into the total cost of ownership is necessary.

Environmental Impact Costs: These are the costs related to the environmental impact of the asset, including its carbon footprint, waste generation, and potential pollution. While these costs may not always directly impact the organization's finances, they are essential considerations in the era of corporate social responsibility and environmental regulations.

By understanding these costs, organizations can make more informed CAPEX decisions and manage their assets more effectively throughout their lifecycle. This comprehensive approach to evaluating the true cost of an asset can also help organizations to better anticipate and budget for future expenses, ultimately leading to improved financial management and business sustainability.

Determining the right maintenance strategy for your physical assets is crucial for efficient asset management. When making this decision, you should base it on an in-depth understanding of your assets, their criticality to your operations, and your organization's specific circumstances. Here are key considerations that will guide you in choosing the most suitable maintenance strategy:

Criticality of the Asset: Not all assets are created equal. Some are more critical to your operations than others. Understanding the role of each asset in your operational processes is the first step to determining the right maintenance strategy. For example, if downtime in a particular asset can lead to significant operational or safety issues, more proactive predictive or condition-based maintenance strategies would be suitable.

Costs and Resources: Maintenance strategies vary in cost implications and resource requirements. Reactive maintenance may have lower upfront costs but can lead to higher costs in the long run due to potential unexpected failures and associated downtime. On the other hand, predictive and cognitive maintenance strategies may require significant investment in technology and expertise. However, they can reduce costs in the long term by minimizing downtime and extending asset lifespan.

Asset Condition and Performance Data: The availability and quality of data about your assets can influence the choice of maintenance strategy. For example, condition-based, predictive, and cognitive maintenance strategies require ongoing data collection and analysis to be effective.

Technological Capabilities: Adopting more advanced maintenance strategies, such as predictive or cognitive maintenance, depends on your organization's technological capabilities. These strategies require sophisticated data analytics, IoT devices for data collection, and sometimes even machine learning algorithms for making predictions.

Regulatory Compliance: In certain industries, regulatory bodies may dictate specific maintenance requirements for certain types of assets. Ensure that your chosen maintenance strategy complies with any such regulations.

Asset Lifecycle Stage: The stage of the asset in its lifecycle also influences the suitable maintenance strategy. For instance, new assets require less frequent maintenance than older ones. A reliability-centered maintenance approach can be helpful in this context, as it considers the asset's age, condition, and operational context.

Risk Tolerance: The level of risk your organization is willing to accept also plays a role. If your organization has a low tolerance for risk, a more proactive and data-driven maintenance strategy would be more suitable.

Choosing the correct maintenance strategy is not a one-size-fits-all decision. It often balances cost, risk, resource availability, and operational requirements. It may also involve combining different asset strategies based on these factors. You should regularly review your maintenance strategies to ensure their effectiveness and alignment with your organizational goals.

When choosing a maintenance strategy for equipment, it's crucial to consider various factors. Below is a general criteria matrix that can help guide this decision:

Criteria	Reactive (Run to Failure)	Preventive	Condition-Based	Predictive	Cognitive
Cost	Low	Medium	Medium-High	High	High
Equipment Age	New/Low Usage	Any Age	Any Age	Mid-Age/Old	Any Age
Equipment Criticality	Low	Medium	High	High	High
Availability of Data	Not Required	Not Required	Required	Required	Required
Technology Requirement	Low	Low	Medium	High	High
Risk Tolerance	High	Medium	Low	Low	Low
Predictability	Low	High	Medium	High	High
Efficiency	Low	Medium	High	High	High

Cost: The overall financial investment required for the strategy. Reactive maintenance has the lowest cost, but the other strategies require more investment due to the need for regular inspections, advanced technology, and data analysis.

Equipment Age: Young or infrequently used equipment might be suitable for a reactive strategy, while any age of equipment can be suitable for other strategies.

Equipment Criticality: For highly critical equipment that cannot afford downtime, condition-based, predictive, or cognitive maintenance strategies are recommended. Less critical equipment might be suitable for a reactive or preventive strategy.

Availability of Data: Condition-based, predictive, and cognitive maintenance strategies require data collection from equipment for effective implementation.

Technology Requirement: The level of technology needed to implement the strategy. Reactive and preventive maintenance requires less technology, while condition-based, predictive, and cognitive maintenance requires more advanced data collection, analysis, and prediction technology.

Risk Tolerance: This relates to the level of risk an organization is willing to tolerate. Reactive maintenance has a high risk of unexpected failures, while the other strategies progressively reduce risk by predicting and preventing failures before they occur.

Predictability: The ability to predict equipment failures. Reactive maintenance has low predictability, while preventive maintenance improves predictability by following a set schedule. Condition-based, predictive, and cognitive maintenance provide the highest level of predictability by using real-time data to predict failures.

Efficiency: The overall efficiency of the maintenance strategy in preventing unplanned downtime and extending equipment life. Reactive maintenance has the lowest efficiency, while the other strategies progressively increase efficiency by preventing failures and optimizing maintenance schedules.

This matrix can serve as a starting point for a wastewater treatment facility when choosing a maintenance strategy. It's important to note that the best strategy may involve a combination of these approaches based on the specific circumstances and needs of the facility.

Understanding the true cost of an asset is vital for devising effective maintenance strategies. The true cost of an asset extends beyond its initial acquisition cost. It encompasses operating, maintaining, downtime, depreciation, disposal, opportunity, risk, and environmental impact costs. By considering these costs, asset managers can better strategize maintenance activities to minimize total asset ownership costs and maximize value.

The True Cost of an Asset

The "Total Cost of Ownership'"(TCO) concept encapsulates all the costs associated with an asset throughout its life cycle. The formula for calculating TCO can be represented as follows:

TCO = Ac + Oc + Mc + Dc + Dp + Dn + Rk + Ec

Where:

- Ac = Acquisition Cost
- Oc = Operating Cost
- Mc = Maintenance Cost
- Dc = Downtime Cost
- Dp = Depreciation
- Dn = Disposal Cost
- Rk = Risk Cost
- Ec = Environmental Impact Cost

The weight of each cost type can vary depending on the asset and the context of its use. For instance, manufacturing equipment's operating cost might carry more weight due to energy consumption. In contrast, the environmental impact cost of a waste disposal system might be more significant due to potential pollution risks.

Impact on Maintenance Strategy

The true cost of an asset can significantly influence the chosen maintenance strategy. A clear understanding of these costs provides insights into where efforts and resources should be focused to maximize asset performance and value. Here's how different cost types could impact maintenance strategies:

Acquisition Cost (Ac): A high acquisition cost might encourage a preventive maintenance strategy to protect the investment and extend the asset's life.

Operating Cost (Oc): If operating costs are high, a predictive maintenance strategy might be ideal to optimize energy use and efficiency.

Maintenance Cost (Mc): When maintenance costs are significant, condition-based maintenance can only be performed when necessary, reducing unnecessary expenditure.

Downtime Cost (Dc): High downtime costs could necessitate a proactive maintenance strategy to prevent unexpected breakdowns and production losses.

Environmental Impact Cost (Ec): A sustainable maintenance strategy that prioritizes eco-friendly practices and materials might be optimal for assets with substantial environmental impact costs.

Each cost consideration should be carefully analyzed to select the most appropriate maintenance strategy for each asset. By integrating the understanding of the true cost of an asset into maintenance planning, organizations can enhance asset performance, reduce costs, and achieve sustainability goals.

[Please note that each cost type's weight and influence on maintenance strategy can be highly context-specific and may require expert advice or consultation with a professional. The provided information is a general guide and should be tailored to each organization's unique needs and circumstances.]

Conclusion

The true cost of an asset plays a pivotal role in shaping asset maintenance strategies. By understanding and considering these costs, asset managers can create strategies that optimize asset value, reduce total ownership costs, and align with environmental sustainability goals. Through scientific calculations and strategic planning, asset maintenance becomes a powerful tool for organizational success and sustainability.

[Note: To ensure accuracy, please consult a professional or specialist in your organization or industry. The formula and costs described above are generalized and might not cover all possible expenses related to your specific assets.]

The Asset Management Phase is a key stage in the overall asset lifecycle. It plays a pivotal role in determining an asset's efficiency, effectiveness, and longevity. This phase encompasses several crucial elements, including regular maintenance, monitoring of asset performance, and decision-making based on collected data and analytics. Here's an overview of the critical steps involved in this phase:

Acquisition: This is the initial stage of asset management, where the necessary assets are identified and procured. During this stage, the organization should consider cost, utility, and asset lifespan factors.

Deployment: After acquiring the assets, you deploy them into the operational environment. This step includes installing and integrating the asset with existing systems and processes. Again, proper planning is necessary to ensure minimal disruption to ongoing operations.

Operation: This is the stage where you put the asset into use. It involves managing the daily operations of the asset to ensure it performs its intended function effectively and efficiently. This phase can also include training personnel to operate the asset correctly and safely.

Maintenance: Regular maintenance is crucial for preserving the condition and performance of the asset. The maintenance schedule can be based on a fixed schedule, condition-based monitoring, or more advanced strategies like predictive or cognitive maintenance. Regular maintenance helps prevent unexpected breakdowns, enhances the asset's lifespan, and ensures optimal performance.

Monitoring and Data Collection: In today's data-driven world, monitoring the performance of assets and collecting data for analysis is vital. IoT devices and advanced data analytics can provide valuable insights into asset performance, potential issues, and areas for improvement.

Analysis and Decision-Making: The data collected from asset monitoring is analyzed to inform decision-making. This can include decisions on maintenance scheduling, replacement of parts, or even asset retirement. Additionally, you can utilize predictive analytics and AI to make more accurate predictions and proactive decisions.

Disposal or Retirement: Eventually, every asset reaches the end of its useful life. At this stage, the asset is decommissioned and disposed of in a manner that complies with regulatory requirements and environmental considerations. The information gathered throughout the asset's lifecycle can inform decisions about the asset's replacement.

You must balance maintaining operational efficiency and managing costs during the Asset Management Phase. An effective asset management strategy ensures that assets deliver maximum value throughout their lifecycle while minimizing costs and risks associated with asset failure.

The Asset Maintenance Phase is a critical stage in the lifecycle of an asset, aimed at preserving and enhancing the asset's functionality, safety, and longevity. Proper asset maintenance can help prevent unexpected breakdowns, reduce operational costs, and extend the asset's useful life. Here's an overview of the essential steps involved in the Asset Maintenance Phase:

Phase 1: Reactive Maintenance; in reactive maintenance, you perform repairs or replacements only after an asset has failed or broken down, making it the most basic form of maintenance. While this approach minimizes upfront maintenance costs, it can lead to higher costs in the long term due to unexpected breakdowns and potentially extensive repairs.

Phase 2: Scheduled Maintenance; Preventive maintenance involves performing regular, scheduled maintenance activities regardless of whether an asset shows signs of failure. The goal is to prevent failures before they happen, reducing the risk of unplanned downtime and costly repairs. You can base maintenance schedules on time intervals, usage, or other operational parameters.

Phase 3: Condition-Based Maintenance; this approach involves continuous or periodic monitoring of an asset's condition and performance to determine when maintenance should be performed. It aims to perform maintenance before potential failure is expected, thus avoiding unnecessary maintenance and reducing downtime. This strategy often utilizes sensors and other monitoring technologies to collect real-time asset condition data.

Phase 4: Predictive Maintenance; building upon condition-based maintenance; predictive maintenance uses advanced analytics, machine learning, and AI technologies to predict when an asset will likely fail or require maintenance. This approach allows for more precise maintenance scheduling and reduces unnecessary maintenance and associated costs.

Phase 5: Cognitive Maintenance; this is the most advanced form of maintenance, where cognitive computing technologies are used to predict potential failures and recommend optimal maintenance actions. Cognitive systems can learn from past data, understand complex patterns, and make highly accurate predictions and recommendations.

Each of these maintenance strategies has its advantages and disadvantages. The choice of strategy depends on various factors, including the type of asset, its operational importance, its failure patterns, and the costs associated with its failure. An effective asset maintenance strategy optimizes the balance between maintenance costs, asset performance, and asset failure risks and impacts.

Exploring Maintenance Strategies through Computerized Maintenance Management System (CMMS)

Maintenance software or Computerized Maintenance Management Systems (CMMS) can be invaluable in forecasting preventive maintenance schedules and balancing resource distribution. These systems can provide visibility into asset conditions, track maintenance tasks, and generate predictive analytics to optimize scheduling. Furthermore, they can help allocate resources, ensuring that work orders are evenly distributed and that all preventive maintenance tasks are adequately staffed.

Preventive Maintenance

Preventive maintenance is a critical component of effective asset management. Regular maintenance activities can extend the lifespan of assets, increase operational efficiency, and prevent costly breakdowns. However, the challenge lies in balancing the distribution of maintenance tasks to ensure a steady workload for maintenance staff. Uneven distribution can lead to periods of excessive work followed by periods of inactivity. This chapter explores strategies for forecasting preventive maintenance schedules and distributing resources evenly.

Preventive Maintenance Scheduling Features

A Computerized Maintenance Management System (CMMS) offers a range of powerful features to facilitate effective preventive maintenance scheduling. Here's how the preventive maintenance schedule feature typically works in a CMMS:

Setting Up Maintenance Schedules: A CMMS allows users to create preventive maintenance (PM) schedules for each piece of equipment in the asset register. These schedules can be defined based on various parameters such as time intervals (e.g., every month), usage intervals (e.g., after every 100 hours of operation), or triggered by specific events or readings (e.g., after a particular meter reading).

Automatic Work Order Creation: Once the preventive maintenance schedules are set up, the CMMS can automatically generate work orders when a maintenance task is due. This feature ensures that no maintenance task is overlooked, thus helping to prevent equipment breakdowns and extend asset lifespan.

Resource Allocation and Task Assignment: The CMMS can also facilitate resource allocation for each preventive maintenance task. This includes assigning tasks to specific technicians or teams, estimating the time and materials required, and scheduling the task at a time that minimizes disruption to operations.

Maintenance History Tracking: A CMMS keeps a detailed record of all preventive maintenance tasks performed on each asset. This information can be helpful for auditing and compliance purposes, as well as for analyzing equipment performance and identifying trends or recurring issues.

Predictive Maintenance: Some advanced CMMS solutions also incorporate predictive maintenance capabilities. By analyzing historical maintenance data and real-time equipment condition data, these systems can predict when a piece of equipment is likely to fail, allowing maintenance tasks to be scheduled proactively to prevent failure.

Cost Tracking and Budgeting:
A significant feature in many CMMS platforms is the ability to track and manage costs associated with preventive maintenance work orders. This is an important aspect of asset management as it directly impacts an organization's financial performance and operational efficiency. Here's how the cost-tracking and budgeting feature typically works in a CMMS:

- **Direct Costs:** For each preventive maintenance work order, the CMMS can record direct costs such as labor (based on hourly rates and time taken), materials used, and any special tools or equipment required.
- **Indirect Costs:** In addition to direct costs, the CMMS can also track indirect costs, such as the impact of downtime during maintenance (e.g., loss of production), administrative costs, and any costs associated with outsourcing or subcontracting.
- **Comparison with Budget:** The CMMS allows for establishing a budget for each preventive maintenance task. It can then compare the actual costs against the budget, providing a real-time view of budget adherence.
- **Historical Cost Data:** The system stores historical cost data for each asset and maintenance task. This information can be used for future budgeting, forecasting, and decision-making regarding the replacement or refurbishment of assets.
- **Cost Reports:** CMMS can generate detailed cost reports, providing insights into the most and least expensive assets to maintain, cost trends over time, cost per unit of output, and other key cost metrics.

Reports and Analytics: Finally, a CMMS can generate various reports related to preventive maintenance. For instance, it can provide an overview of all upcoming maintenance tasks, track the completion rate of preventive maintenance tasks, calculate the cost of preventive maintenance, and more. These reports can provide valuable insights into the preventive maintenance program's effectiveness and help identify improvement areas.

A CMMS can help organizations streamline their preventive maintenance processes, improve equipment reliability, and achieve significant cost savings through these features.

The PM scheduling module in a CMMS is a proactive tool that allows for creating, assigning, and tracking preventive maintenance tasks. It is designed to generate work orders based on pre-set intervals, which could be time-based (e.g., every two weeks), usage-based (e.g., after 100 hours of operation), or condition-based (triggered by a specific parameter, such as temperature or vibration readings).

Let's illustrate this with an example. Consider a manufacturing company with several production lines with different types of machinery. Each piece of equipment may have different maintenance needs and schedules. A conveyor belt might require a check-up every month, while a high-performance engine may need an oil change after every 100 hours of operation. Using a CMMS, the maintenance manager can input these requirements into the system, setting the specific parameters for each piece of equipment. Once set, the CMMS will automatically generate a PM work order when the maintenance task is due. This work order can contain all necessary details, such as task description, required tools and parts, safety instructions, and the estimated task duration.

Preventive maintenance activities are usually planned according to an asset's expected lifecycle or the manufacturer's recommendations. These activities are scheduled with a lead time, last start date, interval, and next start date to launch preventive work orders.

Preventive maintenance schedule forecasting involves predicting when these tasks will need to be performed in the future. This can be done using historical data, manufacturer's guidelines, and asset condition monitoring. The goal is to create a schedule that maximizes asset performance and longevity while minimizing disruptions to operations.

Resource Management and Scheduling Forecast through PM Schedules
Computerized Maintenance Management Systems (CMMS) provide a comprehensive view of all maintenance activities, including the ability to track different tasks related to an asset. When a pump, for example, needs maintenance, it isn't just a single task performed by one craft; it often involves collaboration between different crafts, such as electrical, mechanical, and instrumentation.

A well-utilized CMMS can schedule these different crafts to work concurrently or sequentially, depending on the nature of the tasks, to minimize downtime. This approach ensures maximum resource utilization, with each craft completing its tasks coordinated, providing cost-effective maintenance and minimal disruption to operations.

A crucial aspect of efficient asset management lies in the effective distribution of resources. This includes the workforce and time, materials, and budget allocation for each preventive maintenance task. A Computerized Maintenance Management System (CMMS) facilitates this process by utilizing each asset's Preventive Maintenance (PM) schedules.

A PM schedule is a structured plan that outlines the periodic maintenance required for each asset within an organization. This schedule, linked to each asset in the CMMS, includes details such as the type of maintenance tasks, their frequency, the estimated time required, and the necessary resources and materials.

Here's how a CMMS leverages these PM schedules to facilitate schedule forecasting and resource management:

Schedule Forecasting: The CMMS uses the information in the PM schedules to create a comprehensive forecast of future maintenance activities. The forecasting feature lets the maintenance team view all upcoming tasks for a specific date range. This is accomplished by taking the frequency parameter from each PM schedule and projecting when the next maintenance task will be due.

For example, if a pump requires maintenance every three months, the CMMS will automatically populate the forecast schedule with the next maintenance date three months from the last service date. This forecast allows the team to anticipate future maintenance needs and plan accordingly.

Consider the example of a manufacturing plant with a range of machinery. Each machine would have distinct maintenance requirements and schedules. A cooling system may need servicing every six months, while a milling machine might require calibration after a certain number of operating hours.

The CMMS configures these specifics as PM Schedules for each asset type. Once established, the CMMS uses these PM Schedules to forecast maintenance activities. The forecast view, often referred to as the 'look-ahead' feature, allows maintenance managers to see all the maintenance tasks scheduled for a particular date range. This feature pulls information from the PM Schedules and calculates when the next task is due based on the frequency or usage parameters set in the PM Schedule.

For instance, if the PM Schedule for the cooling system specifies a check-up every six months, the 'look-ahead' feature will display this task in the forecast view six months from the last maintenance date. Suppose the milling machine needs calibration after every 200 hours of operation. In that case, the system will calculate when this threshold will be reached based on the machine's usage rate and display the task accordingly in the forecast view.

This ability to forecast maintenance schedules based on PM Schedules enables maintenance managers to have a comprehensive view of future tasks. With this information, they can plan resources more effectively, ensuring that all maintenance tasks are promptly addressed, and operations run smoothly.

Resource Distribution: Resource allocation is vital to planning, and this is where a CMMS truly shines. By factoring in the details from the PM schedules, such as the estimated time and materials for each task, a CMMS can help distribute resources evenly.
For instance, if a specific maintenance task requires two technicians and is estimated to last three hours, the CMMS considers these factors when assigning work orders. This ensures that resources are not overloaded on a particular day and that enough staff and materials are available for each task. This level of foresight helps to avoid bottlenecks, reduces downtime, and ensures a more streamlined workflow.

Resource Management: The CMMS can manage resources more effectively by using the information in the PM schedules. This includes tracking the availability of technicians, monitoring the use of materials, and even managing the budget for maintenance tasks. A CMMS provides real-time insights into resource usage through comprehensive reports and dashboards. This can help identify trends, such as recurring shortages of specific materials or high demand for particular skills, enabling the organization to address these issues proactively. In summary, a CMMS utilizes PM schedules to streamline the process of schedule forecasting and resource management. This leads to more efficient operations, better use of resources, and ultimately, extends the lifespan of the assets. The next chapter will explore how technological advancements, such as artificial intelligence and machine learning, can further enhance these processes.

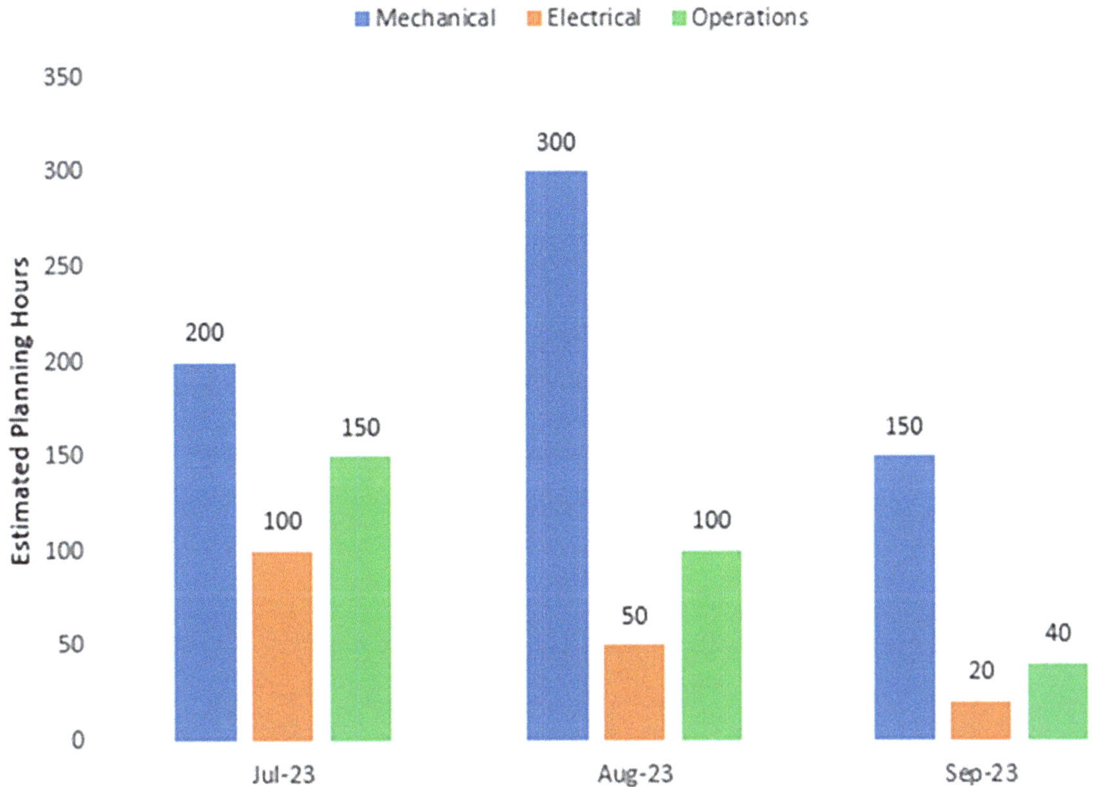

In the bar chart depicted above, it is evident that there is an uneven distribution of estimated planning hours for June, July, August, and September in the third week. The chart illustrates peaks and troughs that indicate a significant variance in the allocation of planning hours over these months.

For instance, in the 1st week of August, there is a considerable surge in the mechanical planning hours, which dips in the following week.

This uneven distribution could suggest a lack of balance in resource allocation or inconsistencies in scheduling. It might lead to periods of overwork during peak times, followed by under-utilization of resources during the lulls. Therefore, reviewing the planning strategy could be beneficial for a more balanced workload distribution, ensuring optimal resource utilization and efficient operations.

Balancing resource distribution is a challenge when scheduling preventive maintenance. The goal is to create an even distribution of work so maintenance personnel are consistently utilized and not overwhelmed by sudden surges of work orders.

There are several strategies to achieve this balance:

Staggering Maintenance Schedules: Instead of performing all preventive maintenance tasks simultaneously, they can be staggered across different periods. This can spread out the workload and make it more manageable.

Flexible Scheduling: While sticking to preventive maintenance schedules is important, there is usually some flexibility. Suppose resources are tight during a particular period. In that case, some maintenance tasks may be rescheduled later when resources are more readily available.

Prioritizing Tasks: Not all preventive maintenance tasks are equally critical. By prioritizing tasks based on the importance of the asset to operations, potential risks of failure, and the complexity of the task, resources can be allocated more effectively.

Cross-Training Personnel: By cross-training maintenance personnel, they can be more flexible in the tasks they perform. This can help balance the workload as personnel can be allocated to different tasks.

In any maintenance operation, especially within a comprehensive Computerized Maintenance Management System (CMMS), it is critical to acknowledge the interdependence of various crafts. For instance, when a piece of equipment, such as a pump, requires maintenance, the tasks typically extend beyond the scope of a single craft. Instead, the operation is a collaborative effort that engages various crafts—mechanical, electrical, instrumentation, and others.

To maximize efficiency and minimize downtime, a well-implemented CMMS can schedule these diverse crafts to work concurrently or sequentially based on the specific nature of the tasks. Such an approach optimizes resource utilization and ensures seamless coordination among the crafts. This leads to cost-effective maintenance with minimal operational disruption.

Implementing Multi-Craft Coordination Using a CMMS: A Step-by-Step Process

Here is a step-by-step process for implementing a multi-craft coordination program using a CMMS, with an example of a pump at a wastewater utility facility:

Step 1: Asset Identification
Identify the asset that requires maintenance. In this scenario, we'll use a pump in a wastewater utility facility as an example.

Step 2: Craft Requirement Analysis
Determine the types of crafts needed to maintain the identified asset. This might include mechanical, electrical, and instrumentation crafts for a pump.

Step 3: Maintenance Task Definition
Define the specific tasks each craft will need to perform as part of the maintenance process. For instance, the mechanical craft might be responsible for dismantling and inspecting the pump, the electrical craft for checking electrical connections, and the instrumentation craft for verifying the readings from the pump's sensors.

Step 4: CMMS Scheduling
Enter the identified tasks into the CMMS. The tasks should be assigned to a preventive maintenance schedule if the maintenance tasks are routine and predictable or to a reactive work order if the tasks are in response to an unforeseen issue.

Step 5: Craft Assignment
Assign the tasks to the corresponding crafts in the CMMS. You can specify whether the tasks should be completed concurrently (at the same time) or sequentially (one after another) depending on the nature of the tasks and the requirements of the maintenance process.

Step 6: Notification and Scheduling

Use the CMMS to notify the relevant crafts of their assignments. The CMMS should provide a schedule for each craft, outlining when their tasks should start and end.

Step 7: Task Execution

Each craft begins work according to the schedule provided by the CMMS. They can use the system to update their tasks' progress and provide any necessary feedback or reports.

Step 8: Monitoring and Coordination

Use the CMMS to monitor the tasks' progress and coordinate between the crafts. Suppose a particular task is taking longer than expected. In that case, the CMMS can reschedule subsequent tasks and notify the relevant crafts of the changes.

Step 9: Task Completion and Verification

Once a task is completed, the relevant craft uses the CMMS to mark it as completed. The system verifies task completion before marking the overall work order as completed.

Step 10: Review and Reporting

After completing the maintenance work, use the CMMS to review the process and generate reports. This can provide valuable insights for improving future maintenance processes and enhancing multi-craft coordination efficiency.

By following these steps, organizations can effectively coordinate multi-craft maintenance tasks using a CMMS, leading to more efficient and cost-effective maintenance processes.

A Computerized Maintenance Management System (CMMS) is a versatile tool that aids in comprehensive work order planning across an organization. Providing a 360-degree view of the maintenance workload allows for more effective planning, leading to improved efficiency and productivity.

Comprehensive Work Order Planning: A CMMS streamlines the work order process by consolidating all relevant information in one place. It records details such as the type of maintenance required, the assets involved, the necessary materials, and the estimated time for completion.

The data allows for a more structured approach to work order planning. With a CMMS, maintenance managers can prioritize tasks based on their urgency or importance, assign them to the appropriate personnel, and even schedule them for a specific time slot.

360-Degree View of Workload: One of the most valuable features of a CMMS is its ability to provide a holistic view of the maintenance workload through an interactive dashboard. This dashboard displays all active work orders, their status, the assets involved, and the resources allocated to each task.

Such a visual representation of the workload helps managers quickly identify any bottlenecks or areas of concern. They can see at a glance if too many tasks are scheduled for a specific time or if a particular asset requires frequent maintenance.

Resource Management and Adjustment: A CMMS allows for real-time tracking of resources, which is crucial for effective work order planning. It monitors the availability of personnel and materials, ensuring that resources are evenly distributed and that no single task is overloaded. Moreover, a CMMS is a dynamic tool that allows for adjustments. Managers can easily update the work order and reallocate resources if a task takes longer than expected or requires more resources.

Streamlined Workload Management: A CMMS allows organizations to systematize their workload management by compiling all pertinent data in a centralized location. This includes information about the nature of the maintenance task, involved assets, required materials, and projected completion time.
Utilizing this data, maintenance managers can methodically plan work orders, prioritize them based on urgency or significance, delegate them to the relevant crafts, and schedule them efficiently. This streamlined approach to workload management increases productivity and prevents task overload.

Effective Interdepartmental Coordination: One of the key benefits of a CMMS is its capacity to enhance coordination between different departments or crafts within an organization. For instance, a maintenance task may require input from both the mechanical and electrical departments. A CMMS enables these crafts to collaborate seamlessly, with each department aware of what the other is doing, minimizing overlaps and redundancies.
In addition, the CMMS can help coordinate with operations to ensure minimal disruption to the production process. This cross-functional collaboration results in more efficient maintenance procedures and contributes to overall organizational effectiveness.

Real-Time Status Updates: A CMMS offers the significant advantage of providing real-time updates about the status of maintenance tasks. This feature is not just beneficial for the maintenance department, but for the entire organization.

Through an interactive dashboard, a CMMS displays all ongoing work orders, their current status, involved assets, and allocated resources. This real-time overview enables quick identification of potential issues, allowing managers to react swiftly and adjust plans accordingly.

Review and Optimization: A CMMS also facilitates the review process by providing detailed reports on completed work orders. Managers can analyze these reports to identify trends or areas for improvement. For instance, if a specific type of maintenance task consistently takes longer than estimated, it might indicate a need for additional training or a revision of the process.

A CMMS equips maintenance and operations managers with the tools they need to plan, manage, and review work orders more effectively. Providing a comprehensive view of the workload and facilitating real-time resource management enables organizations to optimize their maintenance operations and achieve greater efficiency.

Developing In-house Solutions for Schedule Forecasting and Resource Distribution in the Absence of CMMS Support

Some CMMS may not have built-in features for schedule forecasting and resource distribution. This lack of functionality can pose a significant challenge, especially for organizations that rely heavily on efficient scheduling and resource allocation for preventive maintenance activities. However, even without these built-in features, it is still possible to develop these capabilities in-house using available data from your CMMS. This chapter will walk you through how to accomplish this.

Understanding the Foundation: PM Schedule

Before diving into the creation of a forecasting schedule and resource distribution plan, it's vital to understand the importance of the Preventive Maintenance (PM) Schedule. Your PM Schedule is the foundation upon which your forecasting and resource distribution plan will be built. It details when and how often maintenance tasks need to be completed and serves as the starting point for your forecasting.

Transitioning from understanding the basics of work order planning and asset management, we now delve into the practical application of these concepts using some of the most powerful tools available today - Python, Power BI, and Tableau. When utilized effectively, these tools can help an organization predict, visualize, and manage its planned maintenance schedules efficiently and accurately.

In this chapter, we will first explore how to develop a forecasting system using Python, a high-level programming language known for its versatility and efficiency. We will guide you through generating a forecast report by connecting Python to a SQL Server database and executing T-SQL queries.

Next, we turn our attention to Power BI, a business analytics tool by Microsoft that provides interactive visualizations with an interface that is easy to understand, even for lay users. We will discuss creating dynamic visualizations using Power BI by connecting it to a SQL Server database and utilizing stored procedures.

Finally, we will venture into Tableau, a robust data visualization tool that can translate queries into visualizations. We'll guide you through connecting Tableau to a SQL Server database, creating stored procedures, and generating visualizations for your planned maintenance schedules.

By the end of this chapter, you will be equipped with the knowledge to build an in-house forecasting system that can effectively manage workload, coordinate different crafts, and provide real-time status updates across the organization. Let's dive in and explore these powerful tools.

Building a Forecasting System with Python: Connecting to SQL Server and Generating Forecast Reports

```python
import pyodbc
import pandas as pd
from datetime import datetime, timedelta

# Set up the SQL Server connection
conn = pyodbc.connect('DRIVER={ODBC Driver 17 for SQL Server};'
                      'SERVER=server_name;'
                      'DATABASE=database_name;'
                      'UID=user;'
                      'PWD=password')


# Query the PM Schedules data
sql = "SELECT * FROM PMSchedule"
df = pd.read_sql(sql, conn)

# Close the connection
conn.close()

# Convert dates to datetime objects
df['LastExecutionDate'] = pd.to_datetime(df['LastExecutionDate'])
df['NextExecutionDate'] = pd.to_datetime(df['NextExecutionDate'])

# Generate a dataframe for each PM schedule and store them in a list
dfs = []
for _, row in df.iterrows():
    dates = pd.date_range(row['LastExecutionDate'], row['NextExecutionDate'],
freq=row['Frequency']+'D')
    temp_df = pd.DataFrame(dates, columns=['Date'])
    temp_df['PMScheduleID'] = row['PMScheduleID']
    dfs.append(temp_df)

# Concatenate all dataframes in the list
forecast_df = pd.concat(dfs)

# Group by date and count the number of PM Schedules
forecast_df = forecast_df.groupby('Date').count().reset_index()
forecast_df.columns = ['Date', 'Number of PM Schedules']

# Print the forecast data
print(forecast_df)

# Plot the forecast data
```

```python
forecast_df.plot(x='Date', y='Number of PM Schedules', kind='bar')
```

Note: Replace 'PMSchedule' with your actual table name and 'LastExecutionDate', 'NextExecutionDate', and 'Frequency' with your actual column names. Replace 'server_name', 'database_name', 'user', and 'password' with your actual SQL Server details.

This script connects to the SQL Server database, reads the data from the PMSchedule table into a pandas DataFrame, generates the forecast dates based on the PM schedule frequency, and creates a bar chart of the number of PM Schedules per date.

Note: The frequency in the script is assumed to be in days. If your frequency is in weeks or months, you'll need to adjust the freq parameter in the pd.date_range function accordingly. For instance, for a frequency in weeks, you could use freq=str(row['Frequency']*7)+'D'.

Building a Forecasting Report in Tableau

Tableau is a powerful visualization tool but does not directly support generating new data based on complex rules. However, you can generate the PM work order data in a SQL database and then connect Tableau to this database to visualize the data. Here's how you could do it:

- **Create a SQL stored procedure:** This stored procedure would take inputs like start date, end date, and frequency, and output a table of forecasted PM schedules. This stored procedure uses a number table (master..spt_values) to generate the dates for each PM schedule. It calculates how many intervals fit between the last execution date and the end date, then generates that many dates for each PM schedule.

```sql
CREATE PROCEDURE GeneratePMSchedules
  @startDate DATE,
  @endDate DATE
AS
BEGIN
  SELECT PMId, ScheduleDate
  FROM PMSchedule
  CROSS APPLY (
    SELECT DATEADD(DAY, v.number * Frequency, LastExecutionDate) AS
ScheduleDate
    FROM master..spt_values v
    WHERE v.type = 'P'
      AND v.number BETWEEN 0 AND DATEDIFF(DAY, LastExecutionDate, @endDate)
/ Frequency
  ) dates
  WHERE ScheduleDate BETWEEN @startDate AND @endDate
END
```

- **Connect Tableau to your SQL Server database**. In Tableau Desktop, go to "Data" -> "New Data Source" -> "SQL Server". Enter the necessary server information and credentials to establish the connection.
- In the data source tab, in the "Connections" pane, choose the database that contains your stored procedure.
- In the "Tables" pane, navigate to your stored procedure, drag and drop it into the workspace.
- A "Stored Procedure" dialog box will open. Enter your dynamic start and end dates.
- Click "Update Now" or "Automatically Update." Tableau will call the stored procedure and load the data.
- Now you can create a bar chart with the dates on the x-axis and the count of PM schedules on the y-axis.
- Remember, you need to refresh the data source in Tableau whenever the date range changes. This setup allows you to visualize the PM work order forecast in Tableau based on the data generated by the SQL stored procedure.

Building a Forecasting Report in PowerBI

Power BI is primarily a data visualization tool. While it does have data transformation capabilities with Power Query and DAX, it's not typically used to generate new data or run complex simulations like generating forecasted work orders. However, you can use Power BI to visualize the forecasted data after it has been generated.

1. **Create a SQL stored procedure:** This stored procedure would take inputs like start date, end date, and frequency, and output a table of forecasted PM schedules. This stored procedure uses a number table (master..spt_values) to generate the dates for each PM schedule. It calculates how many intervals fit between the last execution date and the end date, then generates that many dates for each PM schedule.

```sql
CREATE PROCEDURE GeneratePMSchedules
  @startDate DATE,
  @endDate DATE
AS
BEGIN
  SELECT PMId, ScheduleDate
  FROM PMSchedule
  CROSS APPLY (
    SELECT DATEADD(DAY, v.number * Frequency, LastExecutionDate) AS
ScheduleDate
    FROM master..spt_values v
    WHERE v.type = 'P'
      AND v.number BETWEEN 0 AND DATEDIFF(DAY, LastExecutionDate, @endDate) /
Frequency
  ) dates
  WHERE ScheduleDate BETWEEN @startDate AND @endDate
END
```

2. In Power BI, click on "Get Data" and then "SQL Server".
3. In the SQL Server database connection window, select "SQL Server", enter your server and database details, and then choose the appropriate authentication method for your database. Click "Connect".
4. In the Navigator window, select "Advanced options" and then enter the SQL command to call your stored procedure with the desired start and end dates.

```sql
EXEC GeneratePMSchedules '2023-06-01', '2023-09-01'
```

5. Click "OK" and then "Load" to load the data into Power BI.
6. Now you can create a bar chart with the dates on the x-axis and the count of PM schedules on the y-axis.
7. Remember that this stored procedure is a simple example and may need to be adapted based on your specific requirements, such as handling meter-based PM schedules. Also, if you need to change the date range, you'll have to refresh the data in Power BI with a new SQL command.

Once you have your forecasting system, the next step is to create a resource distribution plan. This involves mapping out which resources (personnel, equipment, time) will be needed for each forecasted task. The goal is to ensure that resources are allocated efficiently, and tasks are not over or under-resourced.

One approach to this is to categorize maintenance tasks by craft or skill required (e.g., mechanical, electrical, operations) and then map out which tasks require which resources. Once this is done, you can distribute resources accordingly. Remember, the goal here is to meet the requirements of each task and ensure that resources are utilized effectively across all tasks.

Balancing resources within a Preventive Maintenance (PM) Schedule is intricate. The primary goal is to ensure that every task is provided with the required resources while preventing any resource from being underutilized. To illustrate how to manage resource allocation, let's delve into an example using Python, assuming that our Computerized Maintenance Management System (CMMS) comprises three tables: PMSchedule, PMResource, and PMTask.

- **PMSchedule:** This table holds information about the planned maintenance schedule. Key columns typically include the nextExecutionDate and lastExecutionDate, indicating when a particular maintenance task is next due and when it was last performed. The frequency column may indicate how often the task needs to be performed (for example, daily, weekly, monthly), and the estimatedHours column can hold an estimate of how long the task is expected to take.
- **PMResource:** This table contains information about the resources required for maintenance tasks. Resources can include personnel, equipment, materials, and so forth. Each resource record might include the resource's name, type, availability, and any relevant skills or qualifications.
- **PMTask:** This table holds information about the individual maintenance tasks that need to be performed. Each task record might include details such as the task's name, description, the craft or skill required to perform the task, and the associated scheduled maintenance from the PMSchedule table.

Below is a sample Python script that implements a simple resource distribution algorithm. Please note that this is a simplified example, and your real-world application may need adjustments based on your specific business rules and data structure.

The script first merges the necessary data from the PMSchedule, PMResource, and PMTask tables. It then calculates the total estimated weekly hours and determines the average weekly hours. The script then generates a resource distribution plan by moving tasks from over-resourced to under-resourced weeks, aiming to distribute resources as evenly as possible.

```python
import pandas as pd
import pyodbc

# Establish a connection to your SQL Server
conn = pyodbc.connect('Driver={SQL Server};'
                      'Server=your_server_name;'
                      'Database=your_database_name;'
                      'Trusted_Connection=yes;')

# Read the tables into pandas DataFrames
pm_schedule = pd.read_sql_query('SELECT * FROM PMSchedule', conn)
pm_resource = pd.read_sql_query('SELECT * FROM PMResource', conn)
pm_task = pd.read_sql_query('SELECT * FROM PMTask', conn)

# Merge the tables based on their relationships
merged_data = pd.merge(pm_schedule, pm_task, on='taskId')
merged_data = pd.merge(merged_data, pm_resource, on='resourceId')

# Group the data by week and calculate the total estimated hours per week
weekly_hours = merged_data.groupby('week')['estimatedHours'].sum()

# Determine the average hours per week
average_hours = weekly_hours.mean()

# Generate the resource distribution plan
resource_plan = {}
for week, hours in weekly_hours.items():
    if hours > average_hours:
        # If the week is over-resourced, shift some tasks to under-resourced weeks
        under_resourced_weeks = [w for w, h in weekly_hours.items() if h < average_hours]
        if under_resourced_weeks:
            # Move tasks from the over-resourced week to the most under-resourced week
            task_to_move = merged_data[(merged_data['week'] == week) & (merged_data['estimatedHours'] > average_hours)].iloc[0]
            merged_data.loc[task_to_move.name, 'week'] = under_resourced_weeks[0]
    resource_plan[week] = merged_data[merged_data['week'] == week]
```

```
# Now, resource_plan is a dictionary where each key is a week and the value is
a DataFrame of tasks for that week
```

Balancing resources in a maintenance schedule can be a complex task. This is where advanced algorithms and machine learning can play a crucial role.

Advanced algorithms, such as optimization algorithms, can be utilized to distribute resources evenly across all scheduled tasks. They work by analyzing the requirements of each task and the availability and capabilities of each resource. Based on this analysis, the algorithm assigns resources to tasks to optimize a specified objective, such as minimizing total task completion time or maximizing resource utilization.

Machine learning algorithms can enhance this process by learning from past scheduling and execution data. For instance, a machine learning model can be trained on historical data to predict how long a task will take given specific resources. This prediction can then adjust resource allocation and scheduling decisions in real-time.

Moreover, machine learning can uncover patterns and dependencies that might not be immediately apparent. For example, it might find that certain types of tasks tend to take longer when performed by certain resources or that the performance of specific resources tends to deteriorate over time or under certain conditions. Such insights can further refine the resource allocation and scheduling process.

Condition-Based Maintenance

Condition-based maintenance (CbM) is a proactive maintenance strategy that involves monitoring the actual condition of an asset to decide what maintenance needs to be done. This approach dictates that maintenance should only be performed when specific indicators show signs of decreasing performance or impending failure. In the context of a pump in a wastewater treatment facility, let's examine how this can be implemented using Computerized Maintenance Management System (CMMS) software, focusing on vibration data as a key parameter.

Implementing Condition-Based Maintenance for Pumps in a Wastewater Treatment Facility

Step 1: Define Key Condition Indicator

The first step in setting up a CbM program is to define the key condition indicators that signal the health of the equipment. For our pump, we'll consider vibration a significant indicator of its operational status. Vibration levels can tell us a lot about the internal condition of a pump, as increased vibration can be a sign of misalignment, imbalance, or wear and tear within the pump.

Step 2: Establish Baseline and Threshold Values

Next, establish baseline values and threshold limits for the vibration levels. The baseline is the normal vibration level for the pump operating under optimal conditions. Any substantial deviation from this baseline may be a sign of a problem. The threshold value is the maximum allowable vibration level before maintenance action is required. This threshold should be determined based on industry standards, manufacturer's recommendations, and past maintenance data.

For our pump, let's assume that the baseline vibration level, recorded when the pump is operating under optimal conditions, is 2.5 mm/s. This baseline is the normal vibration level for the pump and serves as the reference point for our condition monitoring.
The threshold value is the maximum allowable vibration level before maintenance action is required. This threshold should be determined based on industry standards, manufacturer's recommendations, and past maintenance data. For our example, let's assume we've set a threshold value of 4.5 mm/s. Any vibration level that exceeds this threshold is a cause for concern, indicating that the pump may require maintenance.

Step 3: Continuous Monitoring and Data Collection

Equip the pump with a vibration sensor to monitor and record vibration levels continuously. This data should be automatically fed into the CMMS system for real-time tracking and analysis.

The CMMS software will analyze the incoming vibration data in real-time, comparing the current values against the established baseline and threshold values. If the vibration levels cross the set threshold, the CMMS system will flag this as a potential issue.

For our pump, If the vibration level recorded is 3.0 mm/s, the CMMS system recognizes this as within normal operating range, as it is between our baseline and threshold values.

However, suppose the vibration level rises and crosses our set threshold of 4.5 mm/s. In that case, the CMMS system will flag this as a potential issue, indicating that the pump is operating outside its normal parameters and may require maintenance. This real-time monitoring and analysis can help prevent minor issues from escalating into major failures, thereby enhancing the reliability and longevity of the pump.

Step 5: Maintenance Action

Once the CMMS system identifies a potential issue, it can automatically generate a work order for maintenance to be performed. This work order should include detailed instructions based on the specific issue identified, ensuring that maintenance personnel can swiftly address the problem.

Step 6: Post-Maintenance Review

After the maintenance task is completed, a review should be carried out to assess the effectiveness of the action taken. The pump's vibration levels should return to within normal baseline values. If not, further investigation and possible maintenance actions may be needed. Using a CMMS to implement a condition-based maintenance program, wastewater treatment facilities can ensure that their pumps always operate at optimal performance levels. This strategy not only helps to prevent unexpected failures but also optimizes the use of resources by ensuring that maintenance is only performed when necessary.

Predictive Maintenance

Predictive maintenance is a proactive maintenance strategy that predicts when equipment failure might occur. In this chapter, we will discuss how predictive maintenance can be implemented for a pump in a wastewater treatment facility using a Computerized Maintenance Management System (CMMS) without using machine learning or AI. We will focus on predicting mechanical failures based on flow rate data analysis.

Implementing Predictive Maintenance for Pump in a Wastewater Treatment Facility

Step 1: Understand the Equipment

The first step in implementing predictive maintenance is understanding the equipment and identifying critical parameters that indicate its health status. For a pump in a wastewater treatment facility, one critical parameter could be the flow rate. A pump's flow rate is the fluid volume it can move per unit of time. Significant changes in flow rate can be indicative of an impending mechanical failure.

Step 2: Establish Baseline and Threshold Values

For our pump, let's assume that the baseline flow rate, recorded when the pump is operating under optimal conditions, is 1000 gallons per minute (gpm). This baseline is the normal flow rate for the pump. It serves as the reference point for our predictive maintenance program. The threshold value is the minimum allowable flow rate before maintenance action is required. This threshold should be determined based on industry standards, manufacturer's recommendations, and past maintenance data. Let's assume we've set a threshold value of 800 gpm for our example. Any flow rate that falls below this threshold is a cause for concern, indicating that the pump may require maintenance.

Step 3: Regular Data Collection

The next step is to collect flow rate data from the pump regularly. This could be done using flow meters attached to the pump and connected to the CMMS system. The data should be collected regularly to track any changes in the pump's performance over time.

Step 4: Data Analysis Using CMMS Software

The CMMS software will analyze the incoming flow rate data in real-time. It will compare the current flow rate values against the established baseline and threshold values.
Suppose the flow rate recorded is 950 gpm. The CMMS system recognizes this as within normal operating range, between our baseline and threshold values.

However, if the flow rate drops and crosses our set threshold of 800 gpm, the CMMS system will flag this as a potential issue, indicating that the pump is operating outside its normal parameters and may require maintenance.

Step 5: Initiate Predictive Maintenance Tasks
Based on the analysis, the CMMS system can automatically generate a work order for predictive maintenance when the flow rate falls below the threshold. This allows the maintenance team to address the issue before it leads to a complete failure, minimizing downtime and potentially saving on the cost of more extensive repairs.
By following these steps, wastewater treatment facilities can effectively implement a predictive maintenance program for pumps using flow rate data analysis without the need for complex machine learning or AI algorithms. This predictive maintenance strategy can help increase pump efficiency, prevent unexpected failures, and extend the lifespan of the equipment.

Minimizing costs and operational downtime is paramount in today's highly competitive industrial landscape. Implementing effective maintenance strategies, proper planning, and asset management through a Computerized Maintenance Management System (CMMS) can be crucial in achieving these objectives. This chapter will discuss how these strategies can lead to substantial savings.

Step 1: Implementing an Effective Maintenance Strategy

Organizations can benefit significantly from transitioning from reactive maintenance to a more proactive strategy, such as preventive, predictive, or condition-based maintenance.
For instance, let's assume that a reactive maintenance approach leads to an unexpected breakdown every two months, resulting in 8 hours of downtime each time. This equates to 48 hours of unplanned downtime per year. If the cost of downtime is estimated to be $500 per hour (taking into account lost production, labor costs, etc.), the total cost of downtime amounts to $24,000 per year.

Switching to a predictive maintenance strategy could reduce this downtime by identifying potential issues before they lead to failures. Let's say this approach reduces the frequency of breakdowns to just once per year, resulting in only 8 hours of downtime. This reduces the annual cost of downtime to $4,000, **a savings of $20,000.**

Step 2: Efficient Maintenance Planning with CMMS

Proper planning and scheduling of maintenance tasks can also lead to cost savings. CMMS software can help create efficient maintenance schedules, minimizing time wasted on unnecessary tasks or waiting for resources.
Assume that before implementing a CMMS, technicians spend an average of 2 hours per day on non-productive activities like searching for parts, waiting for instructions, or filling out paperwork. That's 10 hours per week or about 500 hours in a year per technician. If the labor rate is $50 per hour, this inefficiency costs the company **$25,000** per technician yearly. Implementing a CMMS and optimizing maintenance operations reduce these non-productive activities by 50%. This leads to savings of **250 hours per year** or **$12,500 per technician**. For a maintenance team of 10 technicians, the savings amount to $125,000 per year.

Step 3: Asset Management with CMMS

Effective asset management using a CMMS can lead to extended asset life, reduced asset-related costs, and improved asset availability.
For example, consider an asset that costs **$100,000** and has an estimated useful life of 10 years under reactive maintenance. By transitioning to preventive maintenance and tracking the

asset's lifecycle with CMMS, the asset's useful life might be extended to 15 years. This equates to a savings of $100,000 over 15 years, or approximately $6,667 per year.

By incorporating these strategies, an organization can significantly reduce costs and downtime. In our example, the annual savings from improved maintenance strategies, efficient planning, and better asset management total $151,667. This figure illustrates a CMMS and effective maintenance practices' powerful impact on an organization's bottom line.

The wastewater industry relies heavily on various physical assets, such as pumps, treatment plants, and pipelines, to ensure the proper collection, treatment, and disposal of wastewater. Given the critical nature of these assets, it's crucial to adopt the right maintenance strategy to minimize disruptions, maintain service quality, and ensure environmental compliance. Two common strategies this industry uses are "Run to Failure" and "Preventive Maintenance."

Run to Failure: This maintenance strategy allows an asset to operate until it fails before undertaking any maintenance or repairs. While this might seem counterintuitive, it can be a cost-effective approach for non-critical assets where the cost of failure is low and does not significantly impact the overall operation.

Case Study: Sewage Grinder Pump
Sewage grinder pumps were identified as non-critical assets in a small municipal wastewater treatment plant. These pumps were relatively inexpensive and easy to replace, and their failure would not cause significant disruption to the plant's operations or environmental compliance. Therefore, the plant implemented a Run to Failure strategy for these pumps. When a pump failed, it was simply replaced with a new one. Over time, this strategy proved more cost-effective than regular preventive maintenance on these pumps.

Preventive Maintenance: This strategy involves regularly scheduled maintenance tasks to prevent equipment failure. It's typically used for critical assets where failure can lead to significant operational disruptions, high repair or replacement costs, or environmental non-compliance.

Case Study: Wastewater Treatment Plant
In a large-scale wastewater treatment plant, the aeration system, vital for the biological treatment process, was identified as a critical asset. Failure of the aeration system could lead to significant disruptions to the treatment process, potentially resulting in environmental non-compliance and high remediation costs. Therefore, the plant implemented a comprehensive preventive maintenance program for the aeration system. This included regular inspection and cleaning of the aeration tanks, scheduled replacement of diffusers, and routine monitoring of dissolved oxygen levels. This preventive maintenance strategy helped the plant significantly reduce the risk of unexpected failures and consistently comply with environmental regulations.

In conclusion, the choice between Run to Failure and Preventive Maintenance should be based on a thorough understanding of each asset's nature, criticality, and failure patterns, as well as the potential operational, financial, and environmental impacts of asset failure. By adopting the right maintenance strategy, wastewater utilities can optimize their asset performance, reduce operating costs, and ensure the sustainability and reliability of their services.

In the wastewater industry, the sustainability and reliability of operations heavily depend on the condition and performance of various physical assets. Therefore, it's crucial to implement effective maintenance strategies that help avoid unexpected failures and extend asset life. Two such strategies are "Condition-Based Maintenance" and "Predictive Maintenance."

Condition-Based Maintenance: This strategy involves monitoring the actual condition of an asset to decide what maintenance needs to be done. Condition monitoring could include visual inspections, performance data, or other testing methods.

Case Study: Wastewater Pump Station

In a wastewater pump station, pumps are critical assets that can cause significant disruptions if they fail. Therefore, the station implemented a condition-based maintenance strategy using vibration analysis to monitor the pumps' condition. Technicians used handheld vibration meters to measure the vibration levels of the pumps regularly. When the vibration exceeded a pre-defined threshold, indicating potential failure, the pump was scheduled for maintenance. This condition-based maintenance approach helped the pump station avoid unexpected pump failures and extend the pumps' operational life.

Predictive Maintenance: This strategy uses data-driven predictive modeling to predict when an asset will likely fail, allowing maintenance to be scheduled just in time to prevent failure. Predictive maintenance often involves using advanced technologies such as Internet of Things (IoT) sensors and machine learning algorithms.

Case Study: Advanced Wastewater Treatment Plant

In an advanced wastewater treatment plant, the blowers for the aeration process were identified as critical assets. The failure of these blowers could significantly impact the treatment process, leading to potential environmental non-compliance. As a result, the plant decided to implement a predictive maintenance strategy using IoT sensors to monitor parameters such as temperature, pressure, and vibration. This data was fed into a machine-learning model that could predict potential failures based on pattern recognition. Maintenance could be scheduled to prevent the failure when the model predicted an impending failure. This predictive maintenance approach helped the treatment plant reduce downtime, maintain environmental compliance, and optimize maintenance resources.

In summary, Condition-Based Maintenance and Predictive Maintenance represent advanced strategies that can significantly enhance the effectiveness and efficiency of asset management in the wastewater industry. By leveraging these strategies, wastewater utilities can minimize disruptions, optimize maintenance resources, extend asset life, and ensure the sustainability and reliability of their services.

The Role of SCADA in Maintenance

SCADA (Supervisory Control and Data Acquisition) systems play a critical role in asset maintenance, especially in industries with large-scale industrial processes. SCADA systems collect and analyze real-time data, control industrial processes locally or remotely, and directly interact with devices such as sensors, valves, pumps, motors, and more. Below are some ways SCADA impacts asset maintenance:

Real-Time Data Collection and Analysis: SCADA systems gather data from various sensors installed on the equipment. This real-time data collection enables detailed monitoring of equipment parameters and helps identify deviations from standard performance. For example, a temperature sensor on a motor could indicate overheating issues, which might signal a need for maintenance.

Predictive Maintenance: With the help of data analysis and machine learning, SCADA systems can identify patterns and predict potential failures before they occur. This predictive maintenance approach helps prevent unexpected equipment downtime and extends the lifespan of the assets. It also enables businesses to schedule maintenance activities during non-peak hours to minimize disruption.

Asset Performance Optimization: SCADA systems can help optimize asset performance by controlling and adjusting operational parameters based on real-time data. For instance, it can automatically adjust the speed of a motor to maintain optimal efficiency based on the load. This enhances asset performance and reduces wear and tear, minimizing the need for frequent maintenance.

Historical Data Storage: SCADA systems store historical data that can be used to analyze long-term asset performance, predict future performance trends, and plan preventative maintenance schedules. This historical data can also help identify recurring issues and their root causes, leading to more effective maintenance strategies.

Alarm Management: SCADA systems can trigger alarms when operational parameters deviate from predefined norms. These alarms can alert maintenance teams about potential issues, allowing them to take immediate action and prevent further damage.

Integration with Other Systems: SCADA systems can often be integrated with other business systems like Enterprise Asset Management (EAM) or Computerized Maintenance Management Systems (CMMS). This integration allows for a more holistic view of asset performance and maintenance needs, leading to more informed decision-making.

In summary, SCADA systems provide a robust platform for efficient, predictive, and proactive maintenance strategies. By leveraging real-time data, these systems enable organizations to reduce downtime, enhance asset lifespan, and improve overall operational efficiency.

Regulations and Standards in Asset Management

Regulations Overview

Regulations in asset management exist to create a standardized approach towards asset handling, promote safety, increase reliability, and ensure sustainability. These rules and guidelines, which are generally enforced by government agencies or professional bodies, cover various aspects of asset management, including safety protocols, environmental compliance, financial accuracy, and more.

Here's a brief overview of some of the key regulations in the field:

Safety Regulations: Safety regulations aim to reduce the risk of accidents and protect workers' health and safety. They often dictate how assets should be operated, maintained, and disposed of. In the United States, these regulations are typically enforced by the Occupational Safety and Health Administration (OSHA).

Environmental Regulations: These regulations are designed to minimize the environmental impact of asset operation and disposal. For example, assets like vehicles and industrial machinery must comply with emissions standards. Environmental regulations are often enforced by agencies like the Environmental Protection Agency (EPA) in the United States.

Financial Regulations: Financial regulations in asset management are primarily concerned with ensuring the accuracy of financial reporting. These rules often dictate how assets are to be valued and depreciated over time and how these values should be reported. The Sarbanes-Oxley Act (SOX) is a well-known example of financial regulation affecting asset management in the United States.

Industry-Specific Regulations: Depending on the industry, additional regulations may be needed. For example, the pharmaceutical industry must adhere to the regulations set forth by the Food and Drug Administration (FDA), and the aviation industry must follow the Federal Aviation Administration (FAA) guidelines.

International Standards: There are also international standards that, while not strictly regulations, provide widely accepted guidelines for effective asset management. The ISO 55000 series, for instance, sets out principles and requirements for a comprehensive asset management system.

Adherence to these regulations is crucial. Non-compliance can result in penalties and damage to reputation and even affect an organization's ability to operate. However, beyond mere compliance, these regulations also guide organizations toward best practices in asset management, helping them achieve greater efficiency, sustainability, and value from their assets.

The Impact of Regulations on Asset Management

Regulations have a significant impact on how organizations manage their assets. The extent and nature of this impact can vary, depending on the specific industry, the types of assets involved, and the jurisdictions in which an organization operates. Here are some of how regulations can impact asset management:

Compliance Requirements: Regulations often mandate specific procedures for asset management. This could include how assets are maintained, inspected, and decommissioned. For instance, in the utilities sector, regulations may require regular infrastructure inspection to ensure public safety and service reliability.

Record Keeping and Reporting: Many regulations require organizations to maintain comprehensive records of their asset management activities. This can involve keeping detailed logs of maintenance activities, incident reports, asset performance data, and more. Regular reporting to regulatory bodies may also be required.

Risk Management: Regulatory compliance is a significant part of risk management in asset management. Failure to comply with regulations can result in penalties, legal actions, and damage to an organization's reputation. Therefore, asset management strategies often include measures to ensure regulatory compliance, such as regular audits and compliance checks.

Life-Cycle Management: Regulations can influence how assets are managed throughout their life cycle. From acquiring and installing assets through operation and maintenance to their eventual decommissioning and disposal, regulatory requirements must be considered at every stage.

Asset Performance and Efficiency: Some regulations are designed to promote the efficient and sustainable use of assets. For example, energy efficiency regulations may require certain types of equipment to meet specific performance standards. This can influence the types of assets an organization chooses to acquire and how those assets are managed.

Investment Decisions: The regulatory environment can significantly impact investment decisions in assets. Regulatory changes can alter the expected returns from an asset, influencing decisions about whether to invest in new assets, retire old ones, or refurbish existing assets.

In conclusion, the impact of regulations on asset management is broad and significant. Therefore, organizations need to be aware of the regulatory environment in which they operate and ensure their asset management strategies are aligned with regulatory requirements.

Standards in asset management establish a unified and systematic approach to managing assets over their life cycles. These internationally accepted guidelines provide a framework for organizations to achieve their objectives through effective and efficient asset management. Standards also help in maintaining compliance, reducing risk, and improving sustainability. Let's take a look at some significant standards in the asset management field:

ISO 55000 Series: This is an international asset management standard developed by the International Organization for Standardization (ISO). The series consists of three standards:
ISO 55000 provides an overview of asset management and the standard terms and definitions.
ISO 55001 is the requirements specification for an integrated, effective asset management system.
ISO 55002 guides the implementation of such a system.

These standards are designed to help organizations manage the lifecycle of assets, achieve their objectives, comply with regulations, and improve sustainability.

PAS 55: This is a Publicly Available Specification for optimizing physical asset management. It describes the integrated approach required to meet the 'whole life' asset management challenge. Although the ISO 55000 series have superseded it, it is still widely recognized and used.

GFMAM Asset Management Landscape: The Global Forum on Maintenance and Asset Management (GFMAM) developed this document to create a shared understanding of asset management, aligning concepts and terminology internationally.

IAM Anatomy of Asset Management: The Institute of Asset Management (IAM) developed this framework to expand on the GFMAM's Asset Management Landscape, providing additional details and context.

Industry-Specific Standards: Various industries have their standards for managing assets. For instance, the International Civil Aviation Organization (ICAO) has standards for aviation asset management. At the same time, the Federal Energy Regulatory Commission (FERC) sets standards for energy assets in the US.

Adherence to these standards can bring several benefits. For example, it can lead to cost savings by optimizing asset usage and extending asset life, improved risk management, better service delivery, and increased customer satisfaction, and enhanced sustainability and corporate reputation.

The Impact of Standards on Asset Management

Standards play a crucial role in shaping the practices and procedures of asset management in organizations. They establish guidelines, methodologies, and frameworks that help organizations manage their assets effectively and efficiently. Here's how standards impact asset management:

Establishing Best Practices: Standards typically incorporate industry best practices, which can guide organizations in managing their assets effectively. As a result, organizations can improve their efficiency, reliability, and overall performance by aligning their asset management practices with recognized standards.

Enhancing Interoperability: Standards often provide a common language or framework that improves interoperability between systems or components. This is particularly important in asset management, where diverse assets and systems must interact seamlessly.

Improving Quality and Consistency: Standards help to ensure quality and consistency in asset management practices across an organization. In addition, they provide a benchmark against which performance can be measured and evaluated.

Risk Management: By following established standards, organizations can mitigate various risks associated with asset management, such as operational, financial, and reputational risks. Standards provide guidelines for risk identification, assessment, and management.

Facilitating Regulatory Compliance: Many regulatory bodies refer to or require compliance with specific standards. By adhering to these standards, organizations can ensure they meet regulatory requirements, avoiding penalties and legal complications.

Promoting Sustainability: Some standards focus on sustainability in asset management, guiding organizations to manage their assets to minimize environmental impact and encourage social responsibility.

Influencing Investment Decisions: Standards can impact investment decisions in assets. For instance, a standard that promotes energy-efficient equipment might encourage an organization to invest in such assets to achieve compliance and operational cost savings.

Driving Innovation: Standards can also drive innovation in asset management. By defining a clear framework for asset management, standards can encourage organizations to develop new solutions and technologies to meet or exceed those standards.

ISO 55000 is a prime example of a standard series significantly influencing asset management. It provides a comprehensive approach to managing assets over their life cycles, encompassing risk management, cost optimization, quality improvement, regulatory compliance, and sustainability.

Background:

A public utility company operates an extensive network of wastewater treatment plants, sewer systems, and stormwater infrastructure across several states. With an aging infrastructure and stricter environmental regulations, this utility company had to ensure compliance while managing its aging assets efficiently.

Challenge:

The utility company was facing two key challenges:
1. Ensuring regulatory compliance with wastewater and stormwater quality standards set by the Environmental Protection Agency (EPA).
2. Efficient management and maintenance of its ageing infrastructure, including regular inspections and timely repair or replacement of assets.

Solution:

The wastewater utility company implemented a comprehensive asset management program, which combined technology and best management practices to address these challenges. They utilized a GIS (Geographic Information System) to map and track the physical condition of their assets, including pipes, pumps, and treatment plants. This system enabled them to identify assets that were most at risk of failure and prioritize maintenance or replacement efforts.

The company also implemented a real-time monitoring system equipped with IoT sensors at their wastewater treatment plants and in their sewer and stormwater systems. These sensors provided continuous data on wastewater quality parameters like pH, temperature, and pollutant concentration levels. The data was automatically uploaded to a cloud-based system and analyzed using machine learning algorithms to predict potential breaches of regulatory thresholds.

Outcome:

With this technology-enhanced approach, the public utility company improved its regulatory compliance and asset management.

- Regulatory Compliance: The real-time monitoring system allowed the public utility company to proactively identify and address potential non-compliance issues, ensuring they consistently met EPA regulations. It also provided them with data demonstrating compliance during regulatory inspections and audits.

- Proactive Asset Management: Using GIS and IoT technology, the public utility company could proactively identify and address infrastructure at risk of failure. This reduced asset downtime, fewer emergency repairs, and extended lifespan.
- Improved Decision Making: The data collected from the monitoring system and GIS platform provided the public utility company's management with insights for data-driven decision-making regarding infrastructure maintenance, upgrade, and replacement.
- Cost Savings: The utility company achieved significant cost savings by reducing non-compliance penalties and emergency repairs.

This case study demonstrates how a comprehensive asset management approach, integrating technology, can help utilities comply with regulations while managing their infrastructure efficiently.

Climate Action and Sustainability in Asset Management

Climate action and sustainability have become integral to the asset management sector as the world grapples with the mounting impacts of climate change. The potential physical risks of climate change and the transition risks associated with moving towards a low-carbon economy have significant implications for asset management.

Climate action in asset management involves incorporating climate risk and resilience into the asset management process. This includes considering the potential impacts of extreme weather events and longer-term climate trends on asset performance, lifespan, and maintenance needs. It also involves assessing the carbon footprint of assets and finding ways to reduce this through energy efficiency improvements, renewable energy sourcing, and other strategies. Furthermore, climate action in asset management includes investing in assets that support the transition to a low-carbon economy, such as renewable energy infrastructure.

Sustainability in asset management extends beyond climate issues to encompass a broader range of environmental, social, and governance (ESG) factors. Sustainable asset management involves managing assets in a way that generates financial returns and contributes to sustainable development goals. This can include, for example, investing in assets that support clean water and sanitation, reduce inequalities, or promote decent work and economic growth.

In practice, climate action and sustainability in asset management often involve risk management, opportunity identification, and active stewardship. In addition, asset managers may use various strategies and tools, such as ESG integration, sustainability-themed investing, and impact investing, to achieve their climate and sustainability objectives.

A combination of regulatory pressures, investor demands, and recognition of the financial materiality of ESG factors drives the growing focus on climate action and sustainability in asset management. Regulatory bodies worldwide are introducing new requirements for asset managers to disclose how they manage climate and other ESG risks. At the same time, a growing number of investors are seeking to align their investments with their values and sustainability goals.

Incorporating climate action and sustainability into asset management is challenging. These include data gaps, methodological issues, and uncertainties associated with climate scenarios and sustainability metrics. However, asset managers increasingly recognize that these challenges can be overcome and that climate action and sustainability can create value for investors and society.

Asset management and climate action are intrinsically linked. This connection is fueled by the understanding that the financial sector, including asset management, is crucial in addressing climate change. It also recognizes the potential risks and opportunities that climate change presents to the value and performance of assets.

Asset Exposure to Climate Risks

Climate risks can impact the performance and value of assets in two primary ways: physical risks and transition risks. Physical risks arise from the direct effects of climate change, such as extreme weather events and long-term shifts in climate patterns. These can cause physical damage to assets and disrupt their operation. Transition risks arise from the societal and economic shifts towards a low-carbon economy. These can impact the value of assets tied to carbon-intensive sectors and create opportunities for assets tied to renewable energy and other green technologies.

Asset Management's Role in Mitigating Climate Change

Asset managers can contribute to climate action by investing in green assets and divesting from carbon-intensive assets. Green assets can include renewable energy infrastructure, energy-efficient buildings, and companies with strong environmental performance. Divestment from carbon-intensive assets can reduce the financing available to these sectors and encourage them to transition towards more sustainable practices.

Climate-Related Financial Disclosures

Asset managers are increasingly expected to disclose their approach to managing climate risks and contributing to climate action. These disclosures can enable investors to assess the climate resilience of their portfolios and make informed investment decisions. They can also encourage asset managers to strengthen their climate risk management practices and climate action strategies.

Climate Action as a Value Driver

Asset managers are recognizing that climate action can be a significant value driver. Climate-resilient assets can offer enhanced returns and lower risk profiles in a world increasingly shaped by climate change. Climate action can also generate value by aligning asset management with investor expectations, regulatory requirements, and societal demands for a sustainable, low-carbon economy.

Climate Action Strategies in Asset Management

Asset managers can employ various strategies to integrate climate action into their activities. These include ESG integration, which involves incorporating environmental, social, and governance factors into investment decisions; climate scenario analysis, which consists in assessing the potential impacts of different climate scenarios on asset performance; and active ownership, which involves using shareholder rights to encourage companies to take more decisive climate action.

In summary, asset management is crucial in driving climate action and managing climate risks. The connection between asset management and climate action is not just about risk mitigation—it also represents a significant opportunity to create value, drive innovation, and contribute to a sustainable, low-carbon future.

Climate-resilient asset management strategies focus on integrating climate-related risks and opportunities into the asset management process. They enable organizations to safeguard their assets from the physical and transition risks associated with climate change while positioning them to capitalize on the opportunities presented by the shift to a low-carbon economy. Here are some key elements of a climate-resilient asset management strategy:

Climate Risk Assessment

Conducting a comprehensive climate risk assessment is the first step in developing a climate-resilient asset management strategy. This involves identifying the physical and transition risks that climate change could pose to the organization's assets and assessing the potential impacts of these risks on asset performance, value, and lifespan. Tools such as climate risk screening and scenario analysis can be used to understand the potential implications of different climate scenarios and inform asset management decisions.

Asset Adaptation and Resilience Building

Once climate risks have been identified and assessed, the next step is to implement measures to adapt assets and build their resilience to these risks. This could involve physical adaptations, such as enhancing infrastructure resilience to extreme weather events, or operational adaptations, such as diversifying supply chains to reduce climate-related disruption. It could also involve financial adaptations, such as securing insurance coverage for climate risks.

Climate-Smart Investment

Climate-smart investment is a key part of a climate-resilient asset management strategy. This involves prioritizing investment in assets that are resilient to climate risks, can contribute to climate mitigation or adaptation, or can benefit from the transition to a low-carbon economy. It also involves divesting from or reducing exposure to assets that are vulnerable to climate risks or could be devalued by the low-carbon transition.

Stakeholder Engagement and Communication

Stakeholder engagement and communication are crucial for the success of a climate-resilient asset management strategy. This involves engaging stakeholders to understand their views and expectations on climate action, communicating the organization's approach to climate-resilient asset management, and reporting progress and outcomes. Transparent communication about climate risks and actions can enhance stakeholder trust and support and enable informed decision-making.

Continuous Learning and Improvement

Climate change is a complex and evolving issue, and climate-resilient asset management strategies must be flexible and adaptable. This involves regularly reviewing and updating the strategy considering new climate data, risk assessments, and best practices. It also involves learning from experience and continuously improving the organization's approach to climate-resilient asset management.

In summary, climate-resilient asset management strategies involve a proactive and integrated approach to managing climate risks and opportunities. They require a commitment to understanding and responding to climate change and a willingness to innovate and adapt to safeguard assets and seize new opportunities in a changing climate.

Asset management plays a pivotal role in achieving sustainability goals. It acts as a key enabler by ensuring the optimal use of assets throughout their lifecycle, promoting efficiency and limiting waste, and aligning investment decisions with sustainable outcomes. Here are some of the ways asset management can contribute to sustainability:

Optimal Asset Utilization

Asset management's primary purpose is to ensure that assets deliver maximum value for the organization. This includes optimizing the use of assets to reduce waste and inefficiencies, directly contributing to sustainability. In addition, efficient use of assets reduces the demand for new resources and lowers the environmental footprint of operations.

Lifecycle Approach

Asset management adopts a lifecycle approach, considering the impacts of assets from design and acquisition to disposal. This holistic view enables organizations to minimize the environmental impacts of their assets throughout their life, for example, by selecting assets designed for durability, made from sustainable materials, and easily recycled or repurposed at the end of their life.

Sustainable Investment Decisions

Asset management provides a framework for making investment decisions that consider financial returns and environmental, social, and governance (ESG) factors. This enables organizations to prioritize investments in assets contributing to sustainability goals, such as renewable energy infrastructure or energy-efficient equipment.

Risk Management

Asset management also involves assessing and managing risks associated with assets. This includes environmental and social risks, which are increasingly recognized as material factors that can affect the performance and value of assets. By identifying and mitigating these risks, asset management can help to prevent environmental damage and adverse social impacts.

Performance Monitoring and Continuous Improvement

Asset management includes monitoring the performance of assets and seeking opportunities for continuous improvement. This can help to identify and address sustainability issues, such as excessive energy use or high emissions, and drive progress towards sustainability goals.

Stakeholder Engagement

Effective asset management involves engaging with various stakeholders, including investors, customers, employees, regulators, and communities. This can help align asset management with stakeholder expectations and societal values and contribute to broader sustainability outcomes.

In summary, asset management can be key in achieving sustainability goals by promoting efficient and responsible use of assets, guiding sustainable investment decisions, managing environmental and social risks, and engaging with stakeholders to drive sustainable outcomes. By integrating sustainability considerations into asset management, organizations can enhance the performance and value of their assets and make a meaningful contribution to a sustainable future.

Case Study: Sustainable and Climate-Resilient Asset Management at a wastewater utility company.

Background:

A Wastewater utility company that operates in a region highly vulnerable to the impacts of climate change, including sea-level rise and increased frequency of extreme weather events. Recognizing the need to ensure the resilience and sustainability of their infrastructure, they adopted a forward-thinking approach to asset management.

Challenge:

They faced a significant challenge in ensuring that their wastewater treatment plants, pipelines, and pumping stations could withstand climate change impacts while ensuring sustainability in their operations. They needed to integrate climate resilience and sustainability into their asset management strategy.

Solution:

The wastewater utility company launched a comprehensive sustainability and climate-resilient asset management program to address these challenges.

Climate Risk Assessment: They started by conducting a climate risk assessment of their assets, identifying the most vulnerable to climate change impacts.

Resilience Planning: They developed resilience plans for the at-risk assets, including elevating pumping stations, reinforcing treatment plants against flood risk, and installing backup power systems to ensure continued operation during extreme weather events.

Sustainable Asset Management: They committed to a sustainable asset management strategy, which included prioritizing energy efficiency in their operations, utilizing renewable energy sources, and adopting water reuse practices.

Advanced Technology Implementation: The Wastewater utility company utilized technology such as IoT sensors for real-time monitoring of asset conditions and integrated GIS to track asset locations and their associated climate risks.

Outcome:

The company's approach to sustainable and climate-resilient asset management yielded significant benefits:

Improved Climate Resilience: The measures taken enhanced the resilience of this company's infrastructure, reducing disruptions during extreme weather events and ensuring continuity of service.

Enhanced Sustainability: By implementing energy-efficient practices, utilizing renewable energy, and promoting water reuse, they reduced their environmental impact and moved towards a more sustainable operating model.

Cost Savings: Through proactive climate resilience measures, they avoided costly emergency repairs and disruptions, providing long-term cost savings. Sustainable practices also led to operational cost savings through reduced energy consumption.

Regulatory Compliance: The proactive approach taken by the company helped ensure compliance with emerging regulations related to climate change adaptation and sustainability in the wastewater utility industry.

Through this program, this company demonstrates how wastewater utility companies can proactively address the challenges of climate change while enhancing sustainability, offering valuable lessons for the industry at large.

As the global concern for climate change continues to grow, industries are increasingly pressured to reduce their environmental footprint. Asset management, a key operational aspect of many sectors, is no exception to this trend. This section of the book will delve into the intersection of asset management with greenhouse gas (GHG) reduction and carbon credit, two critical components of climate action strategies.

Greenhouse gas reduction involves strategies aimed at reducing emissions of gases such as carbon dioxide (CO_2), methane (CH_4), and nitrous oxide (N_2O) that contribute to the warming of the Earth's atmosphere. In the context of asset management, this can encompass a variety of tactics, from the choice and design of the assets themselves to the operational practices associated with their use, maintenance, and eventual disposal.

On the other hand, carbon credits represent a market-based mechanism designed to incentivize GHG reduction. A carbon credit represents the right to emit a specific amount of greenhouse gases. Companies can earn these credits by undertaking projects that reduce GHG emissions, and they can sell these credits to other companies that need to offset their own emissions.

The role of asset management in these areas is multifaceted. Asset managers have the power to make decisions that directly impact the GHG emissions of their organization. For example, they can choose to invest in energy-efficient equipment or implement maintenance practices that prolong the useful life of assets, thereby reducing the emissions associated with manufacturing new equipment.

Furthermore, asset management can also play a significant role in a company's strategy to acquire carbon credits. By implementing projects that reduce GHG emissions, asset managers can contribute to their organization's stock of carbon credits, which can then be used for compliance with regulations, sold for profit, or bolster the company's reputation for sustainability.

In this section, we will explore these concepts in more depth, delving into the mechanics of carbon credits, the strategies for GHG reduction in asset management, and the opportunities and challenges that these present. We will also look at real-world case studies where asset management strategies have led to GHG reductions and the acquisition of carbon credits. The goal is to equip asset managers with the knowledge they need to turn the challenge of climate change into an opportunity for innovation, efficiency, and value creation.

Carbon credits are a key component of national and international attempts to mitigate the growth in concentrations of greenhouse gases (GHGs). One carbon credit represents the reduction or the sequestration of one metric ton of carbon dioxide or its equivalent in other greenhouse gases. To understand how carbon credits can play a crucial role in asset management, especially in the wastewater utility industry, we first need to understand what they are and how they work.

The Concept of Carbon Credits

Carbon credits, also known as carbon offsets, were conceived under the Kyoto Protocol, an international agreement linked to the United Nations Framework Convention on Climate Change. The idea is to limit the emission of GHGs by providing economic incentives. Businesses or industries that emit less carbon (below their quota) can sell their extra allowances as credits to larger, more polluting industries.

How Carbon Credits Work

The process begins with setting a cap on allowable emissions. Industries that reduce their emissions below this cap can generate carbon credits. Each credit they earn represents one ton of CO2 they have saved from releasing into the atmosphere. These credits can then be traded on the open market, where companies that struggle to reduce their emissions below the cap can purchase them as a more cost-effective way to achieve compliance.

Carbon Credits in the Wastewater Utility Industry

Wastewater treatment facilities are responsible for significant GHG emissions, primarily methane and nitrous oxide. Introducing carbon credits allows these facilities to turn a problem into a potential revenue stream.

Wastewater treatment facilities can earn carbon credits by introducing technologies and processes that reduce GHG emissions. For example, they can invest in more energy-efficient equipment, implement more effective maintenance practices, or even find ways to capture and utilize the methane they produce.

Once these credits are earned, they can be sold on the carbon market, used to offset the facility's own emissions, or even held onto as an investment.

Let's consider a case where a wastewater treatment facility invests in anaerobic digestion technology. This process breaks down the organic matter in the wastewater, producing biogas in the process. This biogas, rich in methane, can be captured and burned to generate electricity.

By doing this, the facility achieves multiple benefits:
- It reduces the amount of methane—a potent GHG—that would have been released into the atmosphere.
- It reduces its reliance on grid electricity (which is often produced from fossil fuels).
- It earns carbon credits in the process.

These credits can then be sold or used, contributing to the facility's financial viability.

While the concept of carbon credits presents a hopeful solution to mitigate the impacts of greenhouse gases on climate change, it has its controversies and challenges. One of the main issues with the carbon credit market is that it is largely voluntary, and there are significant concerns about its effectiveness in achieving a net reduction of greenhouse gases.

The Voluntary Nature of the Carbon Credit Market

The carbon credit market operates on two levels: the compliance market and the voluntary market. The compliance market is regulated by legal agreements such as the Kyoto Protocol, where countries have mandatory emission reduction targets. However, the voluntary market is driven by companies and individuals who buy carbon credits to offset their carbon footprints.

The voluntary nature of this market can be problematic as it relies on the goodwill of companies and individuals rather than enforceable legal obligations. This could lead to a situation where entities with significant emissions choose not to participate in the program, limiting the overall effectiveness of carbon credits in reducing global greenhouse gas emissions.

Effectiveness in Achieving Net Reduction of Greenhouse Gases

Another concern is whether the purchase of carbon credits actually leads to a net reduction in greenhouse gas emissions. Critics argue that because carbon credits allow companies to offset their emissions rather than reduce them, they may simply be enabling these companies to "buy their way out" of their environmental responsibility.

Moreover, there are concerns about the accuracy of carbon offset calculations and the need for more standardization and transparency in evaluating these projects. It is challenging to measure how much of the greenhouse gas emission reduction can be directly attributed to the offset project and how much would have occurred without it.

Additionally, there's a risk of what's known as "leakage," where a carbon offset project in one area might inadvertently cause an increase in emissions in another area. For example, a project that protects a forest to absorb CO_2 may encourage deforestation elsewhere.

Conclusion

While carbon credits present an innovative approach to addressing climate change, it's clear that more work needs to be done to ensure their effectiveness. A more robust, transparent, and regulated carbon credit market might be the key to ensuring that we move towards achieving a net reduction of greenhouse gases. Until then, we must view carbon credits as a part of the solution but not the solution itself. It is one tool in our arsenal in the fight against climate change.

In the era of rising environmental concerns, particularly the escalating threat of climate change, every sector of society is called upon to participate in the collective effort to reduce greenhouse gas emissions. Climate change includes the world of asset management, which is uniquely positioned to make a substantial impact in this area. As stewards of physical infrastructure and equipment, asset managers are responsible and able to incorporate environmentally conscious strategies into their practices.

At its core, asset management is about the lifecycle oversight of physical assets to achieve specific outcomes, often centered on financial or operational performance. However, this definition is expanding in the context of climate change and environmental sustainability. The role of asset managers now extends to the stewardship of assets in a manner that acknowledges and addresses their environmental impact, especially regarding greenhouse gas emissions. This commitment to sustainability can manifest in numerous ways, from the initial design and acquisition of assets, through operation and maintenance, to their eventual disposal or repurposing.

This chapter delves into the role of asset management in reducing greenhouse gas emissions, exploring how this traditionally profit-focused field can and must evolve to meet the challenges of our changing climate. The principles and practices discussed herein illuminate the potential for asset management to be a force for environmental good, demonstrating how every decision to manage assets can contribute to the broader carbon reduction goal. The chapter will shed light on how asset management can influence greenhouse gas emissions and provide insights into asset managers' strategies to foster sustainability and environmental resilience.

As the environmental impact of business operations becomes increasingly critical, the carbon credit market presents opportunities and challenges for asset management. This chapter explores these dynamics, highlighting how carbon credits can be leveraged as strategic tools in sustainable asset management while examining the hurdles that must be navigated.

Opportunities:

Financial Incentives: Carbon credits can provide significant financial incentives for companies. By reducing their carbon emissions, they can generate carbon credits that can be sold on the market, turning environmental stewardship into a potential revenue stream.

Risk Management: Engaging with the carbon credit market can be a key part of an effective risk management strategy. By offsetting their emissions, companies can mitigate their exposure to potential future regulations or penalties related to carbon emissions.

Reputation and Branding: Companies that proactively engage in carbon reduction efforts and participate in the carbon credit market can enhance their reputation as responsible and sustainable organizations. This can be a key differentiator in an increasingly conscious marketplace of environmental issues.

Investor Appeal: Many investors are seeking to invest in environmentally responsible companies. By actively reducing carbon emissions and participating in the carbon credit market, companies can appeal to this growing segment of the investor community.

Challenges:

Market Volatility: The carbon credit market can be volatile, with prices fluctuating based on supply and demand dynamics, regulatory changes, and other factors. This can make it challenging for companies to plan and budget for their carbon reduction efforts.

Complexity and Lack of Standardization: The rules and regulations governing carbon credits can be complex and vary from one jurisdiction to another. This lack of standardization can make it difficult for companies to navigate the carbon credit market.

Verification and Transparency Issues: Ensuring the validity of carbon credits and maintaining transparency in their generation and trading can be challenging. This can lead to skepticism and criticism from stakeholders and potentially undermine the credibility of the carbon credit market.

Potential for 'Greenwashing': There is a risk that some companies may use the carbon credit market as a form of 'greenwashing,' making claims about their environmental performance that are not backed up by real action or improvements. This can damage the reputation of the carbon credit market and undermine genuine efforts to reduce carbon emissions.

In the context of asset management, understanding these opportunities and challenges is essential. By strategically engaging with the carbon credit market, asset managers can contribute to their organization's environmental goals while delivering financial and reputational benefits. However, navigating the complexities and risks of this market requires careful planning, robust systems and processes, and a commitment to transparency and integrity.

Case Study: A Large Wastewater Treatment Plant

A Large Wastewater Treatment Plant in a metropolitan City faced rising operational costs due to high energy consumption and increasing environmental regulations. Recognizing the potential for cost savings and carbon reduction, the plant's asset management team reviewed its operations and infrastructure comprehensively.

Strategy and Implementation

The team identified two key areas for improvement: the aeration process, which accounted for up to 60% of the plant's energy use, and the digestion process, which produced a significant amount of methane, a potent greenhouse gas.

To address these issues, the team implemented a series of changes:

Energy-Efficient Aeration: The plant replaced traditional aeration equipment with more energy-efficient models. These new aerators provided the same oxygenation level with significantly less energy input, reducing the plant's electricity consumption.

Biogas Capture and Utilization: The plant upgraded its anaerobic digesters to capture the methane produced during digestion. This biogas was then used to generate electricity for the plant's operations, reducing its reliance on grid electricity.

Maintenance and Monitoring: The team also implemented a rigorous maintenance and monitoring program to ensure all equipment operated optimally. This involved using predictive maintenance technologies, which enabled the team to identify and address potential issues before they resulted in equipment failures or inefficiencies.

Results and Carbon Credits

These changes led to a significant reduction in the plant's carbon emissions. The plant was able to quantify this reduction and convert it into carbon credits, which were then sold on the voluntary carbon credit market. The revenue from these sales provided an additional financial benefit, offsetting the equipment upgrades and maintenance program costs.

This case study highlights the potential for asset management strategies to lead to carbon credit acquisition. By focusing on energy efficiency and greenhouse gas reduction, the Wastewater Treatment Plant was able to reduce its operational costs, generate revenue through the sale of carbon credits, and contribute to the global effort to combat climate change.

Climate action and sustainability have become central themes in the global discourse, with asset management playing a critical role in these endeavors. While it is easy to be caught in the immediacy of climate action, it is equally important to understand the complexities that underlie these efforts. One such complexity is the holistic consideration of the life cycle of assets, which brings us to Life Cycle Assessment (LCA).

Life Cycle Assessment is a systematic analysis of the environmental impacts of products or systems throughout their life cycle - from the extraction of raw materials, through production and use, to disposal or recycling. It provides a comprehensive view of the environmental trade-offs involved in different decisions, such as replacing traditional equipment with environmentally sustainable alternatives.

However, while environmentally sustainable equipment is always the superior choice, a life cycle assessment may reveal a more nuanced reality. The environmental impact of producing sustainable equipment, or the emissions from its operation, outweighs the benefits of reduced emissions during its use. Conversely, the benefits are even more significant than initially thought when considering the entire life cycle.

This chapter delves deeper into these complexities and explores how life cycle assessments can guide asset management decisions toward truly effective climate action. By understanding the whole picture, we can ensure that our actions today are well-intentioned and well-founded, and beneficial in the long term.

Life Cycle Assessments (LCAs) emerge as an invaluable tool for holistic understanding and effective decision-making as we delve into the complexities of climate action and asset management. But what exactly are LCAs, and why are they so crucial in sustainable asset management?

A Life Cycle Assessment (LCA) is a methodological framework used to evaluate the environmental impacts of all the stages of a product's life, from cradle to grave. This includes extracting raw materials, manufacturing, distribution, use, and disposal or recycling. The LCA approach quantifies the environmental burdens associated with a product, process, or activity by identifying the energy, materials used, and wastes released to the environment.

LCAs offer an objective process to assess the environmental performance of products and processes, allowing for the identification of opportunities to achieve environmental improvements. Conducting an LCA can also inform decision-making in business and policy, helping to prioritize actions based on their potential impact reduction.

In asset management, LCAs are essential in evaluating the environmental impact of different assets over their entire lifecycle. By understanding these impacts, asset managers can make more informed decisions about procurement, maintenance, and disposal, ultimately leading to more sustainable outcomes.

This chapter will explore the concept of LCAs in greater detail, highlighting their importance, methodology, and application in asset management. We will delve into the nuances of LCAs and illustrate how they can inform and enhance sustainability strategies in asset management.

The increasing focus on sustainable practices across industries has emphasized the importance of Life Cycle Assessments (LCAs) in climate action strategies. They are particularly significant in asset management, where assets have wide-ranging environmental impacts that span their entire lifecycle.

By leveraging LCAs, asset managers can comprehensively understand the environmental impacts associated with their assets. It's not just about assessing the direct emissions produced during an asset's operation phase but also considering the emissions associated with the asset's production, transportation, maintenance, and end-of-life disposal.

LCAs systematically evaluate these impacts, offering a fuller picture of an asset's carbon footprint. This in-depth analysis can reveal surprising insights about where the most significant environmental impacts lie, which might not always be in the most obvious places. For instance, a seemingly eco-friendly asset might have a high environmental impact during its manufacturing phase, offsetting the benefits gained during its use phase.

Informed by LCAs, asset managers can make strategic decisions about asset selection, maintenance schedules, and end-of-life management. For instance, they might invest in assets with lower environmental impacts across their lifecycle, even if they are not the cheapest options upfront. Alternatively, they might decide to extend the useful life of an asset through effective maintenance instead of replacing it, thereby minimizing the environmental impact of producing a new asset.

Moreover, LCAs can also inform carbon credit strategies. By quantifying the emissions associated with different assets and processes, asset managers can identify opportunities for carbon reduction that could potentially translate into carbon credits.

LCAs offer a robust and nuanced tool for integrating climate action into asset management. They encourage long-term thinking and sustainability, pushing businesses towards legally compliant and environmentally responsible operations. This chapter will further delve into how LCAs can be practically implemented in asset management, offering a guide for businesses seeking to align their strategies with the principles of climate action.

When conducting Life Cycle Assessments (LCAs), it's crucial to consider the total carbon emissions of an asset over its entire lifespan. This holistic approach is the core principle of LCAs, ensuring that all stages of an asset's life - from raw material extraction to disposal or recycling - are accounted for in the analysis.

Remembering that an asset's carbon footprint extends far beyond its operational phase is important. While this phase can be a significant source of emissions - particularly for energy-intensive assets - other stages can also contribute significantly to an asset's total carbon emissions. For example, the manufacturing stage can be particularly carbon-intensive, involving energy-consuming processes like smelting, welding, and assembly.

Likewise, transporting an asset from the manufacturer to the user can result in substantial carbon emissions, especially when long distances are involved. The disposal or recycling stage can also be a significant source of emissions, particularly when it involves incineration or landfilling.

By considering total carbon emissions in LCAs, asset managers can avoid the pitfall of 'carbon tunnel vision' - focusing solely on one stage of an asset's life to the neglect of others. This comprehensive approach allows for a more accurate understanding of an asset's environmental impact, leading to more informed decisions about asset selection, use, and disposal.

In addition, a focus on total carbon emissions can also reveal opportunities for carbon reduction that might otherwise be overlooked. For instance, by identifying a particularly carbon-intensive stage in an asset's life, asset managers can explore strategies to reduce emissions, such as implementing more efficient manufacturing processes, optimizing logistics, or improving end-of-life management practices.

Considering total carbon emissions in LCAs is key to sustainable asset management. It provides a more realistic picture of an asset's environmental impact and helps guide effective carbon reduction strategies. In the following sections, we'll explore practical steps for incorporating total carbon emissions into LCAs and case studies illustrating how this approach can lead to successful outcomes.

Case Study 1: A Water Utility Company

A Utility Company, the municipal water service provider, was keen on reducing its carbon footprint as part of its sustainability commitment. A key asset in its operations was the fleet of water pumps used in its water treatment plants, which consumed significant energy and had a substantial carbon footprint.

Recognizing the need for a comprehensive perspective, the Utility company conducted a Life Cycle Assessment (LCA) on their pump fleet. The LCA examined carbon emissions across all life stages - from the extraction and processing of raw materials used in manufacturing the pumps, through their operational life, to their eventual disposal.

The LCA revealed that while the pumps' operational phase was a major source of emissions, the manufacturing and disposal stages also contributed significantly to the total carbon footprint. This insight prompted the utility company to consider not just the operational efficiency of the pumps but also the manufacturing processes and end-of-life management in their asset management strategy.

As a result, the utility company started exploring options for pumps manufactured using less carbon-intensive processes and materials and those with better end-of-life recyclability. They also began working with manufacturers to establish a take-back program for used pumps, ensuring they were responsibly recycled or disposed of, further reducing the life cycle carbon emissions.

Case Study 2: A Wastewater Utility Company

A Wastewater Utility Company provides wastewater management services. They decided to undertake an LCA for their wastewater treatment plants, focusing on the aeration system. This critical component consumes much of the energy in the wastewater treatment process. The LCA highlighted the significant carbon emissions during the operational phase of the aeration systems, primarily due to the high energy consumption. However, the LCA also revealed that the manufacturing and transportation stages contributed considerably to the total carbon footprint.

With this information, the utility company changed its asset management strategy. They started to consider aeration systems that had better operational energy efficiency and lower manufacturing and transportation carbon footprints. For instance, they began sourcing aeration systems from manufacturers that used renewable energy in their production processes. They were located closer to the treatment plants to reduce transportation emissions. These case studies demonstrate how LCAs can guide asset management decisions to significantly reduce total carbon emissions and move towards more sustainable operations.

As we navigate the complexities of climate action and sustainability in asset management, one critical area to explore is the true environmental impact of sustainable equipment. The prevailing assumption is that equipment marketed as 'sustainable' or 'green' is always superior to its traditional counterparts. However, the reality may be more nuanced, necessitating a deeper comparative analysis.

In this chapter, I delve into the truth about sustainable equipment. We unpack the concept of sustainability as it applies to physical assets and critically examine the environmental trade-offs of choosing sustainable equipment over conventional alternatives.

We challenge the notion that sustainable equipment is always the best choice from an environmental perspective. Instead, we argue that decision-makers need to consider the total environmental impact of equipment over its life cycle, from production and use to disposal. This holistic approach, known as a life cycle assessment (LCA), accurately depicts the actual environmental trade-offs.

Through comparative analysis, I aim to shed light on the complex dynamics of sustainable equipment in asset management. I aim to empower asset managers and other decision-makers with the knowledge to make informed, responsible choices in pursuing climate action and sustainability.

A few common perceptions and assumptions come to mind when we think of sustainable equipment. We often envision machines that consume less energy, produce less waste, and have a smaller carbon footprint. The benefits are clear: choosing sustainable equipment can reduce our environmental impact and contribute positively towards our sustainability goals.

These assumptions are often based on the operational phase of an asset's life. Sustainable equipment is typically designed to be more efficient in its use of resources, thereby reducing emissions and waste during operation. This could be a pump that uses less energy to perform its function, a vehicle that emits less carbon dioxide per mile, or a light bulb that lasts longer and uses less electricity than its conventional counterparts.

Moreover, sustainable equipment often comes with the promise of lower operating costs. Energy-efficient machines can reduce energy bills, while durable equipment can reduce maintenance and replacement costs. In this way, sustainable equipment is often seen as a win-win situation: good for the environment and the balance sheet.

Environmental regulations and incentives reinforce this positive perception of sustainable equipment. Many jurisdictions offer tax breaks or subsidies for businesses that invest in sustainable equipment, further encouraging their adoption.

However, while these benefits are significant and well-documented, they only represent part of the picture. To fully understand the environmental impact of sustainable equipment, we need to consider its entire life cycle, not just its operational phase. This requires us to challenge our assumptions and take a more nuanced view of what sustainability truly means in asset management.

While the benefits of sustainable equipment in the operational phase are substantial, a comprehensive assessment requires us to look beyond this stage. A life cycle assessment (LCA) evaluates a product's or system's environmental impact over its entire life cycle, from raw material extraction and manufacturing to operation, maintenance, and end-of-life disposal or recycling. This holistic approach allows us to understand sustainable equipment's true environmental costs and benefits.

One of the key insights from LCAs is the concept of environmental trade-offs. For instance, while an energy-efficient pump may consume less energy during its operation, it may require more energy and resources to manufacture and transport than a conventional pump. Similarly, a long-lasting light bulb may use less electricity over its lifetime, but making it could be more carbon-intensive.

Sustainable equipment often incorporates advanced materials or technologies, which can be resource-intensive. The extraction of these materials can result in significant emissions, while the manufacturing processes can be complex and energy-intensive. These upstream impacts are often overlooked when focusing solely on the operational phase.

Furthermore, the end-of-life phase can also present environmental challenges. Sustainable equipment may be more difficult to recycle due to the complex mix of materials used, or it may produce hazardous waste that requires special handling.

These trade-offs do not necessarily negate the benefits of sustainable equipment, but they do complicate the picture. They highlight the importance of taking a life cycle perspective when assessing the environmental impact of our asset management decisions.

Moreover, these trade-offs can have implications for carbon credits. Suppose the upstream and downstream emissions of sustainable equipment are not considered. In that case, businesses may overestimate the carbon savings and the number of carbon credits they can claim. This underscores the need for accurate and comprehensive LCAs to support carbon credit calculations and ensure we are genuinely reducing our greenhouse gas emissions.

In conclusion, while sustainable equipment reduces our environmental impact, we must be mindful of the net benefits and environmental trade-offs. By adopting a life cycle perspective, we can make more informed asset management decisions and contribute more effectively to our sustainability goals.

Life Cycle Assessments (LCAs) provide a valuable tool for understanding the full environmental impact of a product or system. They offer a comprehensive analysis of a product's life from the cradle to the grave, beginning with the extraction of raw materials and manufacturing processes, then proceeding through the product's use and final disposal or recycling. This holistic approach allows us to compare the total carbon emissions of different products or systems, providing a more accurate representation of their environmental impact.

In asset management, LCAs help us make informed decisions about the equipment we invest in. For example, let's consider two conventional and one energy-efficient pump. The energy-efficient pump may have higher upfront costs and consume more resources during manufacturing, resulting in more initial emissions. However, it consumes less energy during its operational phase, resulting in fewer emissions over time.

By conducting an LCA, we can compare the total carbon emissions over the life cycle of the energy-efficient pump versus the conventional pump. If the energy-efficient pump results in lower emissions throughout its life cycle, it can be considered superior from a carbon footprint perspective, regardless of the higher initial emissions.

LCAs are not just limited to energy use and greenhouse gas emissions. They can also consider other environmental impacts, such as water use, air pollution, and waste production. This comprehensive view helps us understand the broader environmental trade-offs involved in our asset management decisions.

It's worth noting that LCAs can be complex and require detailed data about all stages of a product's life cycle. However, they are increasingly supported by robust methodologies and software tools, making them more accessible for businesses of all sizes. By integrating LCAs into our asset management strategies, we can ensure that our efforts to reduce greenhouse gas emissions are genuinely effective and based on solid evidence.

In conclusion, LCAs are vital for making true comparisons between different asset management options. They enable us to see beyond the immediate operational benefits of energy-efficient or renewable technologies and consider the total environmental impact over the product's life cycle. As such, they are essential to our toolkit for sustainable and climate-resilient asset management.

Case Study 1: Energy Industry

In the energy sector, a company aimed to transition from coal-fired power plants to renewable wind farms. A Life Cycle Assessment (LCA) was conducted to analyze the overall environmental impact. The analysis considered every stage, from manufacturing and installation of the wind turbines to their operation, maintenance, and eventual decommissioning. The results showed that, over their lifetime, wind turbines produced significantly less greenhouse gas emissions compared to coal-fired power plants, even when considering the emissions from manufacturing and installation. This case demonstrated the successful implementation of a more sustainable asset, significantly reducing greenhouse gas emissions.

However, the LCA also highlighted an unintended consequence: the wind turbines substantially impacted local bird populations. This biodiversity issue was not initially considered but became crucial to the company's asset management and environmental stewardship strategies.

Case Study 2: Manufacturing Industry

In the manufacturing sector, a company decided to replace its fleet of conventional forklifts with electric ones, aiming to reduce greenhouse gas emissions. An LCA was conducted to compare the total carbon emissions over the life cycle of the electric forklifts versus the conventional ones. The results showed that, while the electric forklifts had higher upfront emissions due to the production of batteries, they resulted in lower emissions throughout their operational life, resulting in net benefits over their lifecycle.

However, the LCA also revealed an unintended consequence: disposing of and recycling the used batteries from the electric forklifts posed a significant environmental challenge. This issue led the company to invest in better battery disposal and recycling programs, ensuring that their shift towards electric forklifts was truly sustainable.

Case Study 3: Transportation Industry

In the transportation sector, a company made a significant investment in electric buses to replace its aging fleet of diesel buses. An LCA was conducted to evaluate the total carbon emissions over the life cycle of electric buses compared to diesel buses. The results showed that, while the electric buses had higher emissions during the manufacturing phase, primarily due to the production of batteries, they emitted significantly less during the operational phase.

However, the LCA also revealed an unintended consequence: the electric buses required more frequent maintenance due to the complexity of their electric drivetrain. This increased maintenance resulted in additional costs and resource use that were not initially considered. The company had to adjust its maintenance and operational strategies to address these challenges and ensure the long-term sustainability of its electric bus fleet.

These case studies illustrate how LCAs can guide asset management decisions toward true environmental sustainability. They highlight the importance of considering the entire life cycle of assets and reveal potential unintended consequences that can inform further strategic planning.

Climate action plans (CAPs) are comprehensive strategies that businesses, governments, and other organizations create and implement to reduce their greenhouse gas emissions and mitigate the impacts of climate change. They are vital tools in the global fight against climate change, as they guide organizations in transitioning towards more sustainable practices, promoting renewable energy, and improving energy efficiency.

A well-designed CAP involves
- setting measurable and achievable goals,
- identifying specific actions to achieve these goals,
- outlining timelines for implementation, and
- establishing mechanisms for monitoring progress and making necessary adjustments.

It can range from simple measures like improving building insulation to more complex strategies like transitioning to renewable energy sources or implementing carbon capture and storage technologies.

However, creating and implementing a CAP is challenging. The first challenge is often the need for adequate data and understanding of the organization's current carbon footprint. This baseline information makes setting realistic goals and measuring progress effectively easier. Another significant challenge is the need for substantial upfront investment. Considerable initial capital outlay is often required for many emissions-reducing actions, such as installing solar panels or upgrading to energy-efficient equipment. While these actions often pay for themselves over time through energy cost savings, the high upfront costs can be a barrier to implementation, particularly for smaller organizations or those with tight budgets.

Engaging stakeholders, whether employees, customers, or the broader community, can also be challenging. Successful implementation of a CAP requires buy-in from all stakeholders, which means effective communication and engagement strategies are crucial.

Finally, the constantly evolving science and technology related to climate change presents a challenge. As new information becomes available and technologies advance, organizations may need to adjust their CAPs to stay current and effective.

In the following chapters, we will explore these challenges in more detail and discuss strategies for overcoming them to implement effective CAPs successfully.

In an era where climate change profoundly impacts economic, social, and environmental dimensions, asset management companies must integrate climate action plans into their strategies. The primary goal is to mitigate the effects of climate change and transition towards a low-carbon economy.

Firstly, climate action plans are instrumental in fulfilling regulatory requirements and avoiding potential legal ramifications. Various jurisdictions worldwide are implementing laws and regulations that require companies, including asset management firms, to reduce their greenhouse gas emissions and take steps to combat climate change. By integrating a comprehensive climate action plan into their strategy, asset management firms can ensure compliance with these regulations and avoid potential fines and legal issues.

Secondly, investors increasingly demand that asset management companies take responsible environmental actions. With rising awareness of environmental issues and a shift towards sustainable investing, asset management firms demonstrating a clear commitment to climate action are more likely to attract and retain clients. A well-developed climate action plan can be a key differentiator in this respect.

Thirdly, climate action plans help asset management firms manage climate-related risks. Physical risks such as extreme weather events can damage infrastructure and other physical assets. Transition risks, such as policy changes and shifts in market preferences, can affect the value of investments. By incorporating climate action plans, asset managers can better anticipate, assess, and manage these risks, thereby protecting their clients' investments.

Furthermore, incorporating a climate action plan can lead to cost savings in the long run. Energy efficiency and renewable energy use can significantly reduce operating costs. Asset management firms can also explore new investment opportunities in green technologies and businesses likely to prosper in a low-carbon economy.

In conclusion, integrating climate action plans in asset management is not just a matter of environmental responsibility. It is also necessary for compliance, risk management, investor attraction, and long-term cost-effectiveness. It is, therefore, essential for asset management companies to understand the need for these plans and how to implement them effectively.

Creating and implementing a climate action plan is a complex process with several potential challenges. Recognizing and addressing these challenges is key to developing a successful and effective plan. Here are some common challenges faced by asset management companies:

Lack of Clear Regulatory Frameworks: In some regions, the lack of clear regulatory frameworks and guidelines can make it difficult for companies to understand their responsibilities regarding climate action. This ambiguity can hinder the development and implementation of climate action plans, as firms may need clarification about what steps they need to take or the targets they should aim for.

Limited Resources and Expertise: Developing and implementing a climate action plan requires significant resources and expertise. Companies may need to conduct extensive research, invest in new technologies or infrastructure, and potentially hire or train staff with expertise in climate action and sustainability. Small and medium-sized enterprises (SMEs) may find these requirements particularly challenging.

Financial Constraints: While implementing climate action strategies can lead to cost savings in the long run, there can be substantial upfront costs. These can include the cost of energy-efficient upgrades, investment in renewable energy sources, and the expense of monitoring and reporting on emissions and other environmental impacts. Convincing stakeholders to support these upfront costs can be challenging, particularly if the return on investment (ROI) is long-term.

Data Availability and Quality: Accurate, high-quality data is crucial for monitoring progress and making informed decisions in implementing a climate action plan. However, gathering this data can be time-consuming and complex, and there can be data availability and quality issues.

Stakeholder Engagement: A successful climate action plan requires buy-in from all stakeholders, including employees, investors, and clients. However, engaging these stakeholders and maintaining their support can be challenging, particularly if they have different views on the importance of climate action or the best strategies to pursue.

Balancing Short-term and Long-term Goals: Asset management companies often face pressure to deliver short-term returns, which can conflict with the long-term goals of a climate action plan. Balancing these differing time horizons and expectations can be a significant challenge. Despite these challenges, the benefits of developing and implementing a climate action plan are substantial. By recognizing and addressing these potential obstacles, asset management companies can develop effective strategies contributing to the fight against climate change while delivering value to their clients and stakeholders.

Despite the challenges, there are effective strategies that asset management firms can employ to overcome hurdles and successfully achieve their climate goals. Below are some of these strategies:

Leverage External Expertise: For firms that lack in-house expertise, leveraging external consultants or partnering with environmental organizations can be an effective way to gain the necessary knowledge and skills. These external resources can guide best practices, help set achievable targets, and provide training to staff.

Secure Adequate Financing: Firms can explore various funding sources to overcome financial constraints. These may include government grants or subsidies for sustainable initiatives, green bonds, or partnerships with financial institutions that offer favorable terms for sustainability-focused investments. Demonstrating the long-term cost savings and potential return on investment of sustainable practices can also help secure funding.

Improve Data Collection and Analysis: To address data availability and quality issues, firms can invest in better data collection and analysis tools. This might include advanced metering systems to track energy use, software solutions for monitoring carbon emissions, or partnerships with data analysis firms that specialize in environmental metrics.

Engage Stakeholders: Engaging stakeholders early and often in the process can help secure their buy-in and support. This might involve conducting regular meetings or workshops to educate stakeholders about the importance of climate action and to gather their input on proposed strategies. Demonstrating the business case for sustainable practices – such as potential cost savings and improved reputation – can also help garner support.

Balance Short-term and Long-term Goals: Firms can develop a phased implementation plan to balance short-term and long-term goals. This might involve starting with smaller, more achievable actions that deliver quick wins and working towards longer-term, more ambitious goals. Firms can also highlight the potential long-term risks of inaction on climate change, such as increased costs and regulatory penalties.

Develop a Clear and Comprehensive Climate Action Plan: A well-developed climate action plan can serve as a roadmap for overcoming challenges and achieving climate goals. The plan should clearly outline the firm's climate goals, the strategies to achieve them, and the metrics it will use to track progress. The plan should be regularly reviewed and updated to reflect changing circumstances and new opportunities.

By implementing these strategies, asset management firms can overcome common challenges and progress significantly toward their climate goals. This can benefit the environment and provide significant business benefits, including cost savings, improved stakeholder relations, and a stronger market position.

In the face of global climate change, the asset management industry increasingly recognizes the importance of integrating sustainability practices into their strategies. A critical aspect of this shift towards greener operations is the implementation of quantitative approaches to climate action planning. Quantitative approaches involve collecting, analyzing, and applying scientific data related to greenhouse gas emissions, energy consumption, life cycle assessments, and carbon sequestration. This data-driven methodology offers a solid foundation upon which effective climate action plans can be built and executed.

This chapter will delve into using quantitative approaches in climate action planning within asset management. We'll explore the various types of climate-related data pertinent to asset managers, discuss how they can be gathered and analyzed, and examine how they can inform asset management strategies. Furthermore, we will present real-world examples illustrating the use of these approaches across various industries. Finally, we will address the challenges of implementing these quantitative methods and suggest possible solutions.

In a world increasingly driven by data, quantitative approaches offer the asset management industry a powerful tool for tackling climate change head-on. By understanding and applying these methods, asset managers can improve their assets' environmental performance and contribute to the global effort against climate change. This chapter will provide valuable insights for professionals in the field as they navigate this important area of their work.

Scientific data is pivotal in climate action, especially in asset management. As organizations strive to mitigate their environmental impact, the need for robust, reliable, and accurate data has never been more apparent. Here's why:

Baseline Establishment: Scientific data allows organizations to establish a baseline of their current greenhouse gas emissions, energy use, water consumption, and other environmental impacts. This baseline is crucial for setting realistic, measurable climate goals and tracking progress.

Informed Decision-Making: With solid data, asset managers can make informed decisions about where to invest resources for the greatest environmental benefit. For example, data might show that a particular piece of equipment is a significant source of emissions, signaling a need for upgrades or replacements.

Risk Assessment: Climate-related data can also help organizations assess their environmental risks. For instance, data might reveal vulnerabilities to extreme weather events, leading to strategies to make the organization's assets more resilient.

Regulatory Compliance: Governments around the world are implementing increasingly stringent environmental regulations. By collecting and analyzing relevant data, organizations can ensure they comply with these regulations and avoid potential penalties.

Transparency and Accountability: Sharing environmental data publicly can enhance an organization's reputation, demonstrating to stakeholders that it is serious about its climate commitments. This transparency can also foster accountability, both internally and externally.

Carbon Credit Opportunities: For organizations participating in carbon markets, scientific data is essential for determining how many carbon credits they may be eligible to receive.

Benchmarking and Continuous Improvement: Continuous data collection allows for benchmarking against industry standards or an organization's historical performance. This facilitates continuous improvement, as organizations can see where they are progressing and where more work is needed.

In the context of climate action in asset management, scientific data is not just a tool—it's a necessity. It provides the foundation upon which successful climate strategies are built. Without it, any climate action plan is merely a shot in the dark.

In the face of climate change, asset management is evolving. It is no longer just about managing physical and financial assets; it's about understanding and mitigating the environmental impacts of these assets. To do this effectively, asset managers need data—specifically, data relevant to climate action. This chapter, "Types of Data Relevant to Climate Action in Asset Management," provides an overview of the data types critical in guiding sustainable asset management decisions.

From energy usage and greenhouse gas emissions to life cycle assessments and climate risk data, a wide range of information can inform climate action strategies in asset management. Understanding the significance of each data type, and knowing how to collect, analyze, and apply this data, can help organizations reduce their environmental impact, meet regulatory requirements, and contribute to a sustainable future.

In this chapter, we will delve into the types of data most relevant to climate action in asset management, explore how this data can be used, and consider the challenges and opportunities associated with collecting and utilizing this data. Whether you are an asset manager looking to incorporate sustainability into your strategies, a policy-maker aiming to drive climate action, or simply someone interested in the intersection of asset management and environmental sustainability, this chapter will provide you with valuable insights.

Greenhouse gas (GHG) emissions data is fundamental to understanding an organization's impact on the climate. This data quantifies the total amount of greenhouse gases, such as carbon dioxide (CO_2), methane (CH_4), and nitrous oxide (N_2O) that are emitted directly or indirectly by an organization.

GHG emissions are quantifiable based on the GHG Protocol principles, the most widely used international accounting tool for government and business leaders to understand, quantify, and manage greenhouse gas emissions. It provides the accounting framework for nearly every GHG standard and program worldwide - from the International Standards Organization (ISO) to The Climate Registry - and hundreds of GHG inventories prepared by individual companies.

The GHG Protocol categorizes emissions into three scopes:
- Scope 1: Direct emissions from owned or controlled sources
- Scope 2: Indirect emissions from the generation of purchased energy
- Scope 3: All indirect emissions (not included in scope 2) that occur in the value chain of the reporting company, including both upstream and downstream emissions.

The calculation of GHG emissions typically involves the following steps:

Identify GHG Sources: Determine the sources of GHG emissions within your organization. This could include combustion processes, industrial processes, and even fugitive emissions (unintentional leakage of GHGs).

Collect Activity Data: Activity data refers to the magnitude of human activity resulting in emissions. This could include the amount of fuel combusted, miles driven by vehicles, or electricity consumed.

Determine Emission Factors: Emission factors are coefficients that quantify the emissions or removals per unit of activity. These can be sourced from databases provided by governmental and international bodies.

Calculate Emissions: Multiply the activity data by the emission factor to calculate the emissions for each source.

For example, the formula to calculate CO2 emissions from the combustion of natural gas can be expressed as:

```
CO2 Emissions = Volume of Natural Gas Combusted (m3) x Emission Factor (kg
CO2/m3)
```

Emission factors can vary depending on the specific characteristics of the fuel being combusted and the combustion technology used, so it's important to use the most accurate and relevant emission factors.

The GHG Protocol provides tools for calculating GHG emissions, including sector-specific guidance for industries like energy, buildings, and transportation. Additionally, many countries provide national GHG inventory reporting guidelines and tools which can also be used to calculate GHG emissions.

Research in this area involves
- developing more accurate and detailed emission factors,
- improving methods for collecting and verifying activity data, and
- developing new technologies and practices to reduce GHG emissions.

Energy consumption data is crucial in managing an organization's carbon footprint and making strategic decisions for climate action. This data measures the amount of energy an organization consumes in its operations. Understanding energy consumption can identify inefficiencies and opportunities for conservation, directly affecting greenhouse gas (GHG) emissions.

The quantification of energy consumption involves capturing data from various sources of energy use, such as electricity, natural gas, and fuel oil, among others.

The calculation of energy consumption generally involves the following steps:

Identify Energy Sources: Determine the different sources of energy consumption within your organization. This can include electricity, heating and cooling (e.g., natural gas, fuel oil), transportation fuels (e.g., gasoline, diesel), and other forms of energy use.

Collect Activity Data: Activity data refers to the quantity of energy consumed. This can be collected from utility bills, fuel purchase records, and meter readings.

Convert Activity Data to Common Units: In order to aggregate energy consumption data across different energy sources, the data must be converted to a common unit of measure. This is typically done in kilowatt-hours (kWh) for electricity or British thermal units (BTUs) for other forms of energy.

For example, the formula to convert natural gas consumption from cubic feet (cf) to BTUs is:

```
Energy Consumption (BTUs) = Volume of Natural Gas Consumed (cf) x Energy
Content of Natural Gas (BTUs/cf)
```

The energy content of natural gas can vary, but the U.S. Energy Information Administration provides an average value of about 1,037 BTUs/cf.

In addition to direct measurements, energy modeling software can estimate energy consumption based on building characteristics, usage patterns, and local climate data. These models can be used to evaluate the potential impact of energy efficiency measures.

Research in this area involves
- developing more accurate methods for measuring and modeling energy consumption,
- exploring new technologies and practices for energy efficiency, and

- understanding the behavioral and organizational factors that influence energy use.

Furthermore, new methodologies are being developed to integrate renewable energy sources and consider the time-varying impacts of energy use on GHG emissions.

Life Cycle Assessment (LCA) data is a comprehensive evaluation of the environmental impacts of all the stages of a product's life from the cradle to the grave. This includes raw material extraction, materials processing, manufacturing, distribution, use, repair and maintenance, and disposal or recycling. LCA is widely recognized as the best tool for environmental decision-making and policy development.

The LCA process is divided into four components:

Goal and Scope Definition: Identify the purpose, the system to be evaluated, the functional unit (i.e., the unit of product system function), and the system boundaries.

Inventory Analysis: Collect data on all inputs (e.g., energy, materials) and outputs (e.g., emissions, waste) within the system boundaries.

Impact Assessment: Assign environmental impact categories and category indicators to the inputs and outputs, and characterize the results.

Interpretation: Evaluate and interpret the results to help make a more informed decision.

The LCA calculations are typically done using software tools, which use databases of life cycle inventory (LCI) data and impact assessment methods. These tools calculate the environmental impacts by multiplying the quantity of each input or output by a factor representing its potential environmental impact.

For example, the Global Warming Potential (GWP) of CO2 emissions is calculated as:

```
GWP (CO2 equivalents) = Mass of CO2 Emitted (kg) x GWP Factor for CO2 (kg CO2
equivalents/kg CO2)
```

The GWP factor for CO2 is defined as 1. The GWP factors are much higher due to their more significant warming effect per unit mass for other greenhouse gases, such as methane (CH4) and nitrous oxide (N2O).

Research in LCA involves improving the accuracy and completeness of LCI data, developing better methods for impact assessment, and exploring new applications of LCA in areas such as sustainable design, policy making, and consumer behavior. There's also ongoing work to incorporate additional dimensions of sustainability (e.g., social, economic) into the LCA framework.

Carbon sequestration and offset data are crucial in climate action and asset management. Carbon sequestration refers to the process of capturing and storing atmospheric carbon dioxide. It is one method of reducing the amount of carbon dioxide in the atmosphere to mitigate global climate change. Carbon offsets are a form of trade. When you buy an offset, you fund projects that reduce greenhouse gas emissions.

Carbon Sequestration Data
Carbon sequestration can occur in two primary ways: biological and geological. Biological sequestration includes photosynthesis in which plants absorb carbon dioxide, release oxygen, and store carbon in their tissues and the soil. Geological sequestration involves injecting carbon dioxide directly into underground reservoirs.
In both cases, a key scientific calculation is the amount of CO2 sequestered per unit of time or unit of area. This can vary widely depending on the specific sequestration method and local conditions. For example, a mature forest can sequester more carbon per hectare per year than a young forest or grassland.

A simple formula to estimate biological carbon sequestration is:

```
Carbon sequestered (kg) = Area (ha) x Sequestration rate (kg CO2/ha/year) x
Time (years)
```

Carbon Offset Data
A carbon offset represents a reduction of greenhouse gases equivalent to one metric ton of carbon dioxide (CO2e). Carbon offset projects can take many forms, such as renewable energy projects, energy efficiency improvements, methane capture from landfills, reforestation, and sustainable agriculture.

The calculation of carbon offsets involves estimating the amount of greenhouse gas emissions that the project avoids or reduces, typically compared to a baseline scenario representing what would have occurred without the project.

A simplified formula to calculate carbon offsets is:

```
Carbon offsets (tonnes CO2e) = Activity data (e.g., kWh of renewable energy) x
Emission factor (tonnes CO2e/unit of activity)
```

Emission factors represent a given source's average greenhouse gas emission rate relative to activity units. They are usually derived from scientific research and are published by reputable sources like the Intergovernmental Panel on Climate Change (IPCC).

Research in carbon sequestration and offsets involves
- improving the accuracy of sequestration rates and emission factors,
- developing better methods for monitoring and verifying the results, and
- exploring new types of offset projects.

Data is the backbone of any effective climate action plan, especially in the context of asset management. However, gathering, analyzing, and interpreting this data can often be challenging. This chapter delves into data collection and analysis, outlining how businesses can effectively gather and utilize data to inform their climate action strategies and drive sustainable asset management.

We'll discuss different methods for data collection, from direct measurements to various monitoring technologies, and explore strategies for ensuring data quality and integrity. Further, we will examine how this data can be analyzed to provide meaningful insights that drive decision-making processes.

This chapter will provide readers with a comprehensive understanding of turning raw data into actionable insights, empowering them to make more informed, sustainable decisions in their asset management practices.

The process of gathering climate-related data is multidimensional. It can be complex, relying on a combination of direct and indirect measurement techniques and various data sources. Here are some of the primary techniques utilized:

Direct Measurement: This method directly measures the variables of interest. For example, greenhouse gas emissions from a particular industrial process could be measured using gas analyzers or sampling systems. Similarly, energy consumption can be measured using meters that track the use of electricity, gas, or other forms of energy.

Indirect Measurement: Sometimes, direct measurement is not possible or practical. In these cases, indirect methods are used. For instance, carbon sequestration could be estimated based on the type and age of a forest and local climate conditions. Similarly, life cycle assessment data often relies on estimates and models to calculate potential environmental impacts over the life of a product or service.

Monitoring Technology: Various technologies can be used to monitor and record climate-related data. For example, sensors and IoT devices can track energy consumption in real-time, providing more granular and up-to-date information than traditional metering methods. In addition, satellite imagery can provide valuable data on land use changes, deforestation, and other factors relevant to carbon sequestration.

Data Sources: There are many sources of climate-related data, ranging from government and international databases to industry reports and scientific studies. These can provide valuable context and benchmarks for a company's data. For example, the Intergovernmental Panel on Climate Change (IPCC) provides comprehensive reports on the state of climate science, including data and models that can be used to estimate greenhouse gas emissions and potential mitigation strategies.

Surveys and Interviews: Surveys and interviews can gather qualitative data on attitudes and behaviors related to climate action. For example, a company might survey its employees to understand their views on sustainability and how it affects their work. This can provide valuable insights that can inform the development of climate action strategies.
In all cases, ensuring the accuracy and reliability of data is critical. This may involve regular calibration of measuring equipment, rigorous data quality checks, and careful consideration of potential sources of error and bias in the data.

Analyzing climate data can be complex, requiring specialized tools and software to handle large data sets, perform statistical analysis, and visualize data in meaningful ways. Here are some of the tools commonly used in this area:

Statistical Analysis Software: Statistical software such as R, Python, SAS, or SPSS can perform a wide range of analyses on climate data. This might include descriptive statistics, trend analysis, correlation and regression analysis, hypothesis testing, and more. These tools can handle large data sets and provide high flexibility and control over the analysis process.

GIS (Geographic Information System) Software: GIS software such as ArcGIS or QGIS can analyze and visualize spatial data. This can be particularly useful when looking at issues like land use changes, deforestation, or the geographical distribution of greenhouse gas emissions.

Life Cycle Assessment (LCA) Tools: LCA tools such as SimaPro, GaBi, or OpenLCA can be used to conduct detailed life cycle assessments, helping to quantify the environmental impacts of products, services, or processes over their entire life cycle. These tools often come with databases of environmental impact data, making it easier to perform these complex calculations.

Energy Management Software: Software like EnergyCAP, eSight, or Energy Manager can help organizations track and manage their energy consumption, as well as identify opportunities for energy efficiency improvements.

Carbon Footprint Calculators: Tools like the GHG Protocol's Carbon Footprint Calculator or the EPA's Greenhouse Gas Equivalencies Calculator can estimate an organization's carbon footprint based on various data inputs.

Climate Modeling Software: Advanced software like the Community Earth System Model (CESM) or the Integrated Assessment Modeling Framework (IAMF) can be used to simulate various climate scenarios' impacts, helping inform strategic planning and risk management.

Data Visualization Tools: Tools like Tableau, Microsoft Power BI, or Google Data Studio can help visualize climate data in a way that's easy to understand and communicate. This can be particularly useful when presenting findings to stakeholders or the public.

Regardless of the specific tools used, ensuring that the data analysis is robust, transparent, and aligned with the organization's climate action goals is essential. This often requires a combination of technical expertise, strategic thinking, and a deep understanding of the specific context and needs of the organization.

Interpreting climate data for decision-making is a critical step in the asset management process. It's not enough to gather and analyze data; it must also be interpreted and contextualized to inform strategic decisions. Here's a look at some key aspects of this process:

Understanding the Context: Before interpreting data, it's essential to understand the context in which it exists. This means considering the broader environmental, social, and economic factors that may be influencing the data, as well as the specific goals and priorities of the organization.

Identifying Trends and Patterns: Interpreting climate data often involves identifying trends and patterns within the data. This includes rising or falling emission levels, changes in energy consumption, or shifts in the lifecycle impacts of different assets. Statistical analysis can help identify these trends and determine whether they are statistically significant.

Assessing Impacts and Risks: Interpreting climate data also involves assessing different climate scenarios' potential impacts and risks. This might involve using models to predict future climate conditions, assessing the potential impacts of those conditions on the organization's assets, and evaluating the risks associated with different mitigation and adaptation strategies.

Making Comparisons: Interpreting data often involves making comparisons between different datasets or between actual data and target values. For example, an organization might compare its emissions data to industry benchmarks or compare actual energy consumption to target values in a climate action plan.

Communicating Findings: Finally, interpreting climate data involves effectively communicating the findings to stakeholders. This might involve creating visualizations, reports, or presentations that clearly and effectively convey the key findings and implications of the data.

In all these steps, it's important to approach the interpretation process critically, consider potential sources of bias or error, and ensure that the interpretation is grounded in sound scientific principles. Additionally, it's crucial to remember that the interpretation of climate data should not be a one-time event. However, rather an ongoing process of monitoring, analysis, and adjustment as new data becomes available and circumstances change.

Scientific calculations form the backbone of evidence-based decision-making in asset management, particularly in the context of climate action. These calculations, derived from many data sources, provide valuable insights to guide strategic planning, optimize operations, and facilitate compliance with environmental regulations. This chapter explores the integration of these scientific calculations into the asset management process, elucidating how they contribute to improved sustainability and efficiency.

From quantifying greenhouse gas emissions to analyzing life-cycle impacts, scientific calculations offer a robust, data-driven approach to managing assets. They enable asset managers to make informed decisions aligning with organizational goals and global sustainability targets. However, applying these calculations isn't straightforward—it requires understanding the relevant data, the appropriate computational methodologies, and the context in which these calculations will be used.

By the end of this chapter, you'll better understand how scientific calculations can be applied to asset management and how they can help organizations transition toward more sustainable and resilient operations.

Greenhouse gas (GHG) emissions calculations play a pivotal role in understanding an organization's carbon footprint, i.e., its contribution to global warming and climate change. These calculations primarily focus on three types of gases: carbon dioxide (CO_2), methane (CH_4), and nitrous oxide (N_2O). Each of these gases has a different global warming potential (GWP), which is used to convert emissions of various gases into carbon dioxide equivalent (CO_2e).

The basic formula to calculate CO_2e is:

```
CO2e = Quantity of Emitted Gas x GWP of the Gas
```

For example, methane has a GWP of 27-30 over 100 years, which means that emitting one ton of methane is equivalent to emitting 27-30 tons of CO2. Therefore, if an organization emits 10 tons of methane, it would be equivalent to 270-300 tons of CO2e.

The implications of GHG emissions calculations are profound. They help in:
Benchmarking and Reporting: GHG emissions data is crucial for reporting to regulatory bodies, investors, and other stakeholders. It aids in benchmarking an organization's performance against industry standards and competitors.

Identifying Reduction Opportunities: By identifying the sources and quantities of emissions, organizations can pinpoint areas where emissions reduction is feasible and prioritize their mitigation efforts accordingly.

Setting and Tracking Goals: GHG emissions calculations form the basis for setting climate targets, such as reducing emissions by a certain percentage in a given year. They also allow organizations to track their progress toward these goals.

Carbon Pricing: The quantification of GHG emissions is the first step towards implementing a carbon pricing strategy, where a financial cost is associated with emitting carbon dioxide or other GHGs.

Risk Management: Understanding GHG emissions can help organizations anticipate regulatory, reputational, and physical risks associated with climate change.

Essentially, GHG emissions calculations equip asset managers with the data to make informed decisions, drive sustainability efforts, and contribute to global climate action.

Energy efficiency calculations are a critical component of sustainable asset management. These calculations measure the ratio of output (useful energy or work) to input (energy consumed) in systems, processes, or devices. The better the ratio, the higher the energy efficiency. The fundamental formula for calculating energy efficiency is:

```
Energy Efficiency (%) = (Energy Output / Energy Input) x 100%
```

This ratio helps asset managers understand how well their assets perform regarding energy use. For instance, if a particular machine requires 1000 joules of energy to perform a task and successfully converts 800 joules into useful work, its energy efficiency is 80%.

Energy efficiency calculations play an instrumental role in asset management for several reasons:

Cost Savings: More efficient assets consume less energy, reducing operational costs. By calculating energy efficiency, managers can identify inefficient assets and take steps to improve their performance, such as replacing or upgrading equipment.

Environmental Impact: Energy production and consumption contribute significantly to greenhouse gas emissions. Improving energy efficiency can therefore reduce an organization's carbon footprint and contribute to sustainability goals.

Regulatory Compliance: Energy efficiency standards and regulations are increasingly being introduced globally. Adherence to these standards requires accurate calculations and reporting of energy efficiency.

Asset Lifespan: Energy-efficient assets often have longer lifespans. Energy efficiency can extend the useful life of assets by reducing wear and tear on the equipment, reducing the need for premature replacements and associated costs.

Reputation Management: Consumers and investors are increasingly interested in businesses' environmental practices. Committing to energy efficiency can enhance an organization's reputation and provide a competitive edge.

In conclusion, energy efficiency calculations provide valuable insights that can influence strategic decisions in asset management, from procurement and maintenance to disposal.

These calculations contribute to cost savings and support environmental sustainability and regulatory compliance.

Carbon offset calculations are integral to managing the environmental impact of an organization's operations. Carbon offsets represent a reduction, avoidance, or sequestration of greenhouse gas (GHG) emissions that are used to compensate for emissions produced elsewhere. The basic principle of carbon offsetting is to balance out the GHG emissions you cannot avoid by investing in projects that reduce or remove the equivalent amount of GHGs.

The calculation of carbon offsets often involves the following steps:

Establish Baseline: Determine the GHG emissions your operations produce without any carbon offset projects. This is often calculated in metric tons of CO2 equivalent (CO2e).

Identify Offset Project: Choose a project that will reduce or sequester GHGs. These could include renewable energy projects, forest conservation or reforestation initiatives, and methane capture from landfills.

Calculate Emissions Reduction: Determine the amount of GHGs the offset project will reduce or sequester over its lifetime. This is also measured in metric tons of CO2e.

Calculate Offsets: Subtract the emissions reduction of the offset project from your baseline emissions. The result is your net emissions after accounting for carbon offsets.
The significance of carbon offset calculations in asset management is manifold:

Climate Action: Carbon offset calculations allow businesses to take actionable steps towards reducing their carbon footprint, contributing to the global effort to mitigate climate change.

Regulatory Compliance: Many jurisdictions require businesses to offset a certain amount of their emissions. Accurate carbon offset calculations help ensure compliance with these regulations.

Sustainability Goals: Carbon offset calculations can be instrumental in meeting corporate sustainability targets. They provide a measurable way to track progress toward these goals.

Financial Savings: In some cases, investing in carbon offset projects can lead to financial savings in the long run, especially considering the increasing costs associated with carbon taxes and penalties for exceeding emissions caps.

Stakeholder Engagement: Investors, customers, and employees, are increasingly interested in a company's environmental performance. Committing to carbon offsetting can enhance a company's reputation and brand value.

Climate change poses significant challenges for businesses across all sectors, and asset management is no exception. The implications of climate change and the transition to a low-carbon economy have far-reaching impacts on the value and performance of assets. Therefore, it is essential to incorporate climate data into asset management strategies.

This chapter will discuss the importance of integrating climate data into asset management decisions. We will explore how asset managers can use climate data to assess risks and opportunities associated with their assets and make informed decisions aligning with long-term sustainability goals.

We will explore different methods to integrate climate data into asset management, the role of climate scenario analysis, and how these insights can inform investment decisions. This chapter will also highlight the importance of transparency and disclosure in communicating the climate impacts and risks associated with assets.

By incorporating climate data into asset management strategies, businesses can improve their resilience to climate-related risks, seize opportunities presented by transitioning to a low-carbon economy, and contribute to global efforts to mitigate climate change. The aim is to safeguard assets from climate-related risks and identify and capitalize on opportunities for sustainable growth and innovation.

Greenhouse gas (GHG) emissions data is a fundamental component of climate action in asset management. By understanding the GHG emissions associated with an asset, managers can formulate and implement strategies to reduce these emissions.

Energy Efficiency Measures: A significant proportion of GHG emissions come from energy use. Hence, improving energy efficiency is often a key strategy for reducing emissions. This could involve investing in energy-efficient equipment or improving the efficiency of operations.

Switching to Renewable Energy: Asset managers could consider switching to renewable energy sources. This could involve installing renewable energy systems, such as solar panels or wind turbines, or purchasing green energy from the grid.

Material and Waste Management: Waste production and material use can also contribute to GHG emissions. Asset managers can implement strategies to reduce waste and use materials more efficiently, which can significantly impact GHG emissions.
Carbon Offsetting: While the primary goal should always be to reduce emissions directly, carbon offsetting can be used as a supplementary strategy. This involves investing in projects that reduce or remove GHG emissions elsewhere to offset the emissions that cannot be eliminated.

Green Building Practices: For assets that involve buildings, implementing green building practices can be a significant way to reduce emissions. This could include everything from installing insulation to reduce energy use to incorporating green spaces to absorb CO_2.

Investing in Low-carbon Technologies: New technologies are continually being developed to reduce GHG emissions. By investing in these technologies, asset managers can reduce the emissions associated with their assets and potentially gain a competitive advantage.

These strategies should not be considered in isolation. Instead, the best approach usually involves a combination of strategies tailored to the specific characteristics of the asset and the opportunities available. Regular monitoring and evaluation of GHG emissions data are also crucial to assess the effectiveness of these strategies and make necessary adjustments.

Energy efficiency is critical in managing assets sustainably and addressing the challenges posed by climate change. By adequately analyzing energy consumption data, asset managers can make more informed decisions resulting in environmental and economic benefits.

Identifying Energy Intensive Assets: Energy consumption data allows asset managers to identify which assets consume the most energy. This information can be used to prioritize efforts and resources toward improving the energy efficiency of these assets.

Optimizing Operations: Detailed energy data can help identify operational inefficiencies. For instance, it can highlight peak energy use periods, allowing managers to implement measures to flatten the demand curve, such as demand response strategies or time-of-use energy pricing.

Investment Decisions: Energy efficiency data can inform investment decisions. For example, suppose an asset consistently demonstrates high energy use. In that case, investing in a more energy-efficient replacement may be more cost-effective in the long run.

Benchmarking Performance: Energy consumption data can be used to benchmark the performance of an asset against similar assets. This can provide valuable insights into how well an asset is performing and where there may be opportunities for improvement.

Compliance with Standards and Regulations: Increasingly, businesses must comply with standards and regulations regarding energy efficiency. Regular monitoring of energy consumption data can help ensure that assets remain compliant and avoid potential penalties.

Sustainability Reporting: Many stakeholders, including investors, customers, and regulators, are increasingly interested in a company's environmental performance. Energy efficiency data can contribute to sustainability reporting and demonstrate a company's commitment to responsible environmental management.

Remember, regular data collection and analysis are key to effective energy management. Asset managers can quickly identify and respond to changes by continually monitoring energy consumption, ensuring that assets remain as energy efficient as possible.

Carbon offsetting has emerged as an essential tool for businesses in the quest to mitigate climate change. It allows them to compensate for their greenhouse gas (GHG) emissions by investing in projects that reduce, remove, or sequester GHGs elsewhere. Carbon offset data plays a crucial role in asset management, particularly in the following areas:

Achieving Net-Zero Emissions: Carbon offset data is essential for organizations aiming to achieve net-zero emissions. It provides the numerical basis for calculating how much offsetting is needed after all feasible in-house reductions have been made.

Informed Investment Decisions: Carbon offset data can guide investment decisions by highlighting the effectiveness of various offset projects. Not all offset projects are created equal – they differ in cost, GHG reduction potential, and co-benefits (such as biodiversity conservation or community development). Detailed carbon offset data can help asset managers choose the most beneficial projects.

Compliance with Regulations: Some jurisdictions require businesses to offset a certain amount of their emissions. Carbon offset data is necessary to demonstrate compliance with such regulations.

Risk Management: Climate change poses significant risks to businesses directly (e.g., through physical impacts on assets) and indirectly (e.g., through policy and market changes). By helping to mitigate climate change, carbon offsetting can be a form of risk management. Carbon offset data is crucial for understanding and quantifying this benefit.

Reputation and Stakeholder Relations: Many stakeholders, including customers, employees, and investors, value action on climate change. Carbon offset data can be used in sustainability reporting to communicate an organization's climate actions and enhance its reputation.

Future-Proofing Assets: As the world moves towards a low-carbon economy, assets associated with high levels of GHG emissions may become "stranded" or lose their value. Carbon offset data can inform strategies to future-proof assets against this risk.

However, it's important to remember that carbon offsetting should not be seen as a substitute for reducing emissions directly. It should be part of a comprehensive climate strategy that prioritizes emission reductions and uses offsetting to neutralize residual emissions.

Data Accuracy and Reliability:
One of the main challenges in using quantitative approaches in climate action is ensuring the accuracy and reliability of the data. Inaccurate data can lead to incorrect conclusions and suboptimal decisions. The complexity of climate science and the vast scale of data collection can also contribute to uncertainties.

Solution: Implementing robust data collection and verification methods is essential to ensure data accuracy. Using established protocols and standards for data collection can help reduce errors. Additionally, using software tools with inbuilt data validation features can also aid in improving data reliability.

Data Interpretation:
Interpreting climate data correctly and meaningfully is another significant challenge. It requires a deep understanding of climate science and statistical methods. Misinterpretation of data can lead to incorrect decisions and ineffective climate action strategies.
Solution: Training and capacity building in climate science and data analysis is crucial for correctly interpreting data. Additionally, collaborating with climate scientists and data analysts can provide valuable insights. Advanced data analysis tools that provide clear, interpretable outputs can also be helpful.

Integrating Climate Data into Decision-Making:
Integrating it into decision-making can be challenging even when accurate and reliable climate data is available. This is because climate data is just one of many factors that must be considered in decision-making.
Solution: Developing clear guidelines on how climate data should be used in decision-making can help overcome this challenge. These guidelines could specify, for example, how climate risks should be weighed against other risks or how carbon pricing should be incorporated into financial analysis. Tools that integrate climate data into decision-making processes can also be beneficial.

Scaling and Replicating Successful Approaches:
Even when a quantitative approach is successful in one context, scaling or replicating it in other contexts can be challenging due to differences in data availability, institutional structures, and other factors.
Solution: Building flexibility into quantitative approaches can help them adapt to different contexts. Documenting and sharing successful approaches can also facilitate their replication. Collaborating with other organizations and participating in knowledge-sharing networks can provide learning opportunities from others' experiences.

Keeping Pace with Evolving Science and Technology:
Climate science and data analysis technologies are continually evolving. Keeping pace with these changes can be challenging. However, ensuring that the most recent and reliable data and the most effective analysis methods are essential.

Solution: Continuous learning and professional development are essential to keep up with evolving science and technology. Engaging with scientific and professional communities, attending conferences and workshops, and subscribing to relevant journals and newsletters can help stay up-to-date. Partnerships with research institutions and technology providers can also provide access to the latest knowledge and tools.

Below are case studies from different industries that illustrate the triumphs and trials of implementing climate action plans.

Manufacturing Industry:
In the manufacturing industry, a large multinational corporation set ambitious goals to reduce its carbon footprint by 50% over a decade. The company invested heavily in renewable energy, energy-efficient machinery, and advanced metering systems to track their energy use. They also sought external expertise to train their staff in sustainable practices and implemented a comprehensive climate action plan.

Triumphs: The company reduced its carbon emissions by 40% within the first five years, well ahead of its targets. The investments in energy-efficient machinery also led to cost savings in the long run.

Trials: The company faced significant upfront costs in purchasing new machinery and renewable energy systems. There were also challenges in training staff and changing established work practices.

Real Estate Industry:
A real estate company in the commercial sector decided to incorporate sustainability measures into its portfolio of buildings. The company introduced energy-efficient heating, ventilation, and air conditioning (HVAC) systems, installed solar panels on several properties, and improved insulation to reduce energy consumption.

Triumphs: The improvements significantly reduced energy use, saving the company substantial energy costs. It also enhanced the company's reputation among tenants and the public, increasing property values.

Trials: The company encountered resistance from some tenants concerned about potential disruptions while installing the new systems. It also had to deal with the complexity of regulatory requirements in the various jurisdictions where it owned properties.

Transportation Industry:
A public transportation provider had the objective of transitioning its fleet to electric buses to reduce carbon emissions. The company procured electric buses, installed charging infrastructure, and trained its staff on the new technology.

Triumphs: The transition substantially reduced greenhouse gas emissions and improved air quality in the areas the buses served. Using electric buses also led to cost savings due to lower fuel and maintenance costs.

Trials: The company had to contend with high upfront costs for procuring the buses and installing the charging infrastructure. There were also technical issues with the new buses and charging systems initially, requiring additional investment.

These case studies illustrate that while implementing a climate action plan can be challenging, the benefits often outweigh the trials. With careful planning, stakeholder engagement, and a willingness to invest in sustainable practices, industries can effectively reduce their carbon footprint and contribute to global climate action efforts.

Artificial Intelligence in Asset Maintenance

Introduction

Artificial Intelligence (AI) is a transformative technology that has entered various sectors, including asset maintenance. This field, traditionally reliant on manual inspection and predetermined maintenance schedules, is being redefined by the advent of AI. The introduction of AI in asset maintenance has opened a world of possibilities, leading to more efficient and effective maintenance practices.

AI is a branch of computer science that aims to create systems capable of performing tasks that usually require human intelligence. These tasks include understanding natural language, recognizing patterns, solving problems, and making decisions. In the context of asset maintenance, AI can be used to predict future maintenance needs, detect anomalies, automate inspections, and much more.

Artificial Intelligence (AI), as a concept in the context of asset maintenance, refers to the deployment of advanced computing capabilities to simulate human intelligence in analyzing, predicting, and optimizing asset performance and maintenance procedures. It can revolutionize how industries approach asset management by introducing efficiency, precision, and foresight that far exceed manual capabilities.

At its core, AI is a branch of computer science that aims to create systems capable of performing tasks that ordinarily require human intelligence. These tasks include understanding natural language, recognizing patterns, making decisions, and solving complex problems. AI's capabilities are powered by algorithms that learn from data - a concept known as machine learning. These algorithms can recognize patterns in data and make predictions or decisions without being explicitly programmed to perform the task.

Artificial Intelligence (AI) has a wide array of applications in asset maintenance, helping businesses transform their operations, increase productivity, reduce downtime, and save costs. The following are a few key ways in which AI is currently being applied in the field of asset maintenance:

Predictive Maintenance: One of the most prominent applications of AI in asset maintenance is predictive maintenance. Using machine learning algorithms, AI can analyze vast amounts of data from sensors and other sources to predict when an asset is likely to fail or require maintenance. This allows businesses to schedule maintenance proactively, thereby reducing unexpected downtime and extending the life of their assets.

Anomaly Detection: AI can help identify anomalies or unusual patterns in the performance of assets. By continuously monitoring the operational data from the assets, AI can detect deviations from the normal performance that might indicate a potential issue. This early warning system allows businesses to address problems before they become serious, avoiding costly repairs and downtime.

Automated Inspections: AI can automate the inspection process when combined with technologies like drones or robotics. This is particularly useful for assets located in hard-to-reach or hazardous locations. In addition, AI can analyze the data collected during these inspections to identify potential issues, making the inspection process safer, more efficient, and more effective.

Resource Optimization: AI can also be used to optimize the allocation of resources for maintenance activities. By predicting future maintenance needs, AI can help businesses plan and prioritize their maintenance activities, ensuring that resources are used effectively and that maintenance is performed optimally.

Fault Diagnosis: AI can also aid in diagnosing the cause of asset failures. Using machine learning algorithms, AI can analyze historical data to identify patterns or correlations that might indicate the cause of a failure. This can reduce the time it takes to diagnose and fix issues, minimizing downtime.

Knowledge Capture and Transfer: AI can capture the knowledge and expertise of experienced maintenance personnel, codifying this information so it can be used to train less experienced staff. This can help businesses manage the loss of knowledge that occurs when experienced staff retire or leave.

Cognitive Maintenance: AI can enable cognitive maintenance in a more advanced application. This involves using AI to understand and learn from the historical data of an asset, the environmental conditions, and the maintenance history to provide more accurate and timely maintenance predictions.

In the following sections of the book, I will delve into each of these applications in more detail, providing real-world examples and case studies to illustrate how AI is transforming asset maintenance.

Increased Efficiency: AI can process and analyze vast amounts of data much faster than humans, enabling quicker decision-making and increased operational efficiency.

Predictive Capabilities: Through machine learning, AI can predict potential failures before they occur. This allows for preventative maintenance, reducing downtime, and increasing asset life.

Reduced Costs: By predicting and preventing unnecessary maintenance, AI can help to reduce maintenance costs. Additionally, by increasing efficiency and reducing downtime, AI can lead to further cost savings.

Enhanced Safety: With AI, hazardous or difficult inspections can be automated, reducing the risk to human workers and improving safety.

Improved Decision-Making: AI can provide valuable insights and analytics that inform better asset management and maintenance decision-making.

Data Quality and Availability: AI and machine learning requires large amounts of high-quality data to be effective. The AI models may only perform well if the data is accurate, complete, and consistent.

Technology Integration: Integrating AI technology with existing systems and processes can be challenging and may require significant time and resources.

Skills Gap: Understanding and effectively utilizing AI requires specialized skills that may not be readily available within an organization. As a result, training or hiring new staff may be necessary.

Cost of Implementation: While AI can provide significant benefits, the initial cost of implementation can be high, particularly for small and medium-sized businesses.

Privacy and Security Concerns: With the increased use of AI and data collection, data privacy and security issues have become more prominent and must be addressed carefully.

Understanding these benefits and challenges is crucial for organizations considering the adoption of AI in their asset maintenance strategies. In the following sections, we will explore strategies for overcoming these challenges and maximizing the benefits of AI.

Harnessing the power of artificial intelligence (AI) for predictive and condition-based maintenance can be a game-changer for organizations in various industries. AI-based predictive maintenance relies on machine learning models to analyze historical and real-time data from various sensors on an asset, enabling it to predict potential failures or breakdowns before they occur. This foresight allows for timely maintenance interventions, which can significantly reduce downtime and extend the asset's life.

On the other hand, condition-based maintenance leverages AI to continuously monitor the condition of an asset and schedule maintenance when certain predefined conditions are met. This approach, which is often based on factors like vibration, temperature, or sound patterns, can help identify subtle changes in the asset's performance that may indicate an emerging issue. By combining these two AI-driven approaches, organizations can achieve a more proactive, efficient, cost-effective maintenance strategy.

Introduction to Predictive Maintenance

Predictive maintenance (PdM) is a proactive maintenance strategy that predicts when equipment failure might occur. This approach uses condition-monitoring equipment and advanced analytics to monitor equipment in real-time, allowing maintenance to be scheduled at a convenient time before the equipment fails, thus minimizing disruption to production and extending the lifespan of the equipment.

Predictive maintenance leverages technologies such as machine learning, artificial intelligence (AI), and predictive modeling to analyze the operational data of the equipment, identify patterns, and predict future outcomes based on this data. This can include factors like vibration, temperature, and pressure levels.

Predictive maintenance aims to allow businesses to move from reactive maintenance (fixing things when they break) and scheduled maintenance (replacing parts on a regular schedule, whether needed or not) to a strategy that only performs maintenance when necessary. This can significantly reduce maintenance costs, improve operational efficiency, and prevent unplanned downtime. The main challenge lies in the correct implementation and the requirement of advanced data analysis capabilities. However, with the advancement of technology, predictive maintenance is becoming more accessible and easier to implement.

Artificial Intelligence (AI) plays a pivotal role in predictive maintenance by facilitating the processing and analyzing of vast quantities of data to predict equipment failure. Machine learning algorithms, a subset of AI, can learn from historical data to recognize patterns and anomalies that may indicate future failure.

Artificial Intelligence (AI) has been crucial in predictive maintenance in the wastewater utilities sector. By leveraging data gathered from numerous sensors and applying advanced machine learning algorithms, AI can help predict equipment failures, optimize maintenance schedules, and increase operational efficiency. Here's how:

Predicting Equipment Failures: AI systems can analyze data from sensors installed on various assets such as pumps, valves, and motors. These sensors track temperature, pressure, vibration, and flow rates. The AI system can identify patterns and anomalies in this data using machine learning algorithms, indicating potential equipment failures.
For example, an AI system was implemented at a wastewater utility to monitor pump performance in real-time. The system could detect subtle changes in pump vibration and temperature that were precursors to failure. This allowed the utility to service or replace the pump before it failed, avoiding costly downtime.

Optimizing Maintenance Schedules: AI allows utilities to implement a more efficient condition-based maintenance strategy instead of following a traditional time-based maintenance schedule. The AI system continuously monitors the condition of the equipment. Then, it uses predictive algorithms to estimate when maintenance will be required.
At a public utility company, for instance, an AI system was used to optimize maintenance schedules for wastewater treatment plants. The system used data from plant sensors to predict when maintenance would be required, allowing the utility to service equipment just-in-time, reducing unnecessary maintenance costs and extending equipment lifespan.

Improving Operational Efficiency: AI can also help utilities improve operational efficiency by optimizing processes and reducing energy consumption. AI systems can analyze process data and identify inefficiencies or opportunities for optimization.

For example, a large public utility company uses an AI system to optimize the operation of its wastewater treatment plants. The system analyzes process data and suggests adjustments to aeration rates and chemical dosages, significantly reducing energy use and chemical costs while maintaining or improving effluent quality.

AI can be used in the manufacturing industry to monitor the condition of machinery and predict failures before they occur. An AI system can process data from sensors monitoring machine conditions such as vibration, temperature, and pressure. It then uses machine learning algorithms to learn from the historical data about when these conditions led to failures in the past. If similar patterns are detected, the system can send an alert that maintenance is required, thus preventing a costly breakdown.

One example is General Electric's Predix platform, a cloud-based Industrial Internet of Things (IIoT) operating system. Predix uses machine learning algorithms to analyze data from industrial machines and predict when they will need maintenance.

Airports like Delta and EasyJet have employed AI for predictive maintenance in the aviation industry. They use systems that can process vast amounts of data from aircraft systems and components, identify anomalies, and predict possible failures. This allows airlines to perform maintenance when needed rather than on a set schedule, reducing downtime and costs.

In the energy sector, wind farm operators use AI to predict when wind turbines need maintenance. Systems monitor variables such as wind speed, temperature, and turbine vibration and use machine learning to predict when a turbine is likely to fail. By performing maintenance before this happens, operators can avoid costly downtime and increase the lifespan of the turbines.

These examples highlight how AI can be harnessed to predict equipment failures, optimize maintenance schedules, and ultimately save time and resources. However, the successful application of AI in predictive maintenance requires accurate data, appropriate machine learning models, and the right expertise to interpret the results.

Condition-Based Maintenance (CBM) is a maintenance strategy that involves monitoring the actual condition of an asset to decide what maintenance needs to be done. It revolves around the principle of performing maintenance when specific indicators show signs of decreasing performance or impending failure. This strategy aims to minimize both the likelihood of failure and maintenance costs.

Instead of conducting maintenance on a predefined schedule, maintenance is carried out based on trends identified in data collected from equipment sensors. Indicators such as vibration, temperature, noise, and other physical properties are monitored to determine when maintenance is necessary. As a result, CBM can help to prevent unexpected equipment failures and allows for planned maintenance, thus minimizing downtime and extending the life of the equipment.

In the context of physical asset maintenance, CBM is highly relevant. Assets such as machinery, equipment, and infrastructure can be equipped with sensors that collect data on various aspects of their operation. This data can then be analyzed to identify patterns or signs that the asset may be deteriorating or in danger of failing. By acting on this information, maintenance can be carried out just in time to prevent failure without wasting resources on unnecessary maintenance.

In summary, Condition-Based Maintenance provides a proactive approach to maintenance management. Leveraging the power of data and predictive analytics allows organizations to optimize their maintenance activities and improve overall asset reliability.

Artificial intelligence (AI) plays a pivotal role in Condition-Based Maintenance (CBM) due to its ability to efficiently process vast amounts of data and draw meaningful conclusions. AI technologies, such as machine learning, can analyze data from sensors installed on physical assets to predict potential failures or maintenance needs.

One of the primary ways AI contributes to CBM is through predictive analytics. Machine learning algorithms can be trained on historical data from equipment sensors to identify patterns preceding failure. Once trained, these algorithms can analyze real-time data to predict when a piece of equipment is likely to fail or need maintenance. This allows organizations to intervene proactively, preventing equipment failure and minimizing downtime.

For example, Artificial Intelligence (AI) plays a significant role in enabling this predictive approach, providing valuable insights for maintenance decision-making in wastewater treatment facilities.

Real-Time Equipment Monitoring: AI and IoT technology allows for real-time equipment monitoring. For example, IoT sensors installed on various assets such as pumps, filters, and motors in wastewater treatment plants collect data on temperature, pressure, vibration, flow rates, and more variables. AI algorithms then analyze this data to determine the actual operating condition of the equipment.

For instance, a wastewater treatment facility implemented AI-driven real-time monitoring of its centrifugal pumps. The AI system could identify subtle changes in pump vibrations and temperature levels, detecting issues that might not be noticed during routine inspections.

Predictive Analytics for Maintenance: AI systems can analyze the data collected from the equipment to identify patterns and anomalies using machine learning algorithms. These patterns and insights can be used to predict potential equipment failures, allowing maintenance to be carried out before the failure occurs.

At another wastewater treatment facility, an AI system was employed to analyze data from its belt filter presses. As a result, the system identified patterns indicating an imminent belt failure, enabling the maintenance team to address the issue proactively and prevent an unexpected breakdown.

Optimizing Maintenance Schedules: In a CBM strategy, maintenance is only performed when specific indicators show signs of decreasing performance or impending failure. AI plays a crucial role in analyzing the condition of equipment and predicting when maintenance will be required, thus optimizing maintenance schedules and resources.

For example, a wastewater utility uses an AI system to optimize the maintenance schedules of its wastewater treatment plants. The system continuously monitors the conditions of various assets. It uses machine learning algorithms to estimate when maintenance would be required. This allowed the utility company to transition from a regular time-based maintenance schedule to a condition-based one, leading to significant cost savings and more efficient use of resources.

Enhancing Lifespan of Assets: By identifying potential issues early and allowing timely maintenance, AI-driven CBM can prevent minor issues from becoming major problems, which can lead to catastrophic equipment failure. This not only prevents costly repairs or replacements but also enhances the overall lifespan of the assets.

Another example is those condition monitoring systems equipped with AI capabilities that are used to monitor wind turbines in the wind energy sector. Sensors installed on the turbines collect data on vibration, temperature, and acoustics, among other factors. AI algorithms analyze this data to detect anomalies suggesting an impending failure. As a result, maintenance can be scheduled just in time to prevent a breakdown, thereby increasing the turbine's operational efficiency and lifespan.

Another example can be found in the railway industry. Rail operators use AI-powered CBM systems to monitor the condition of trains and railway infrastructure. AI algorithms analyze data from sensors installed on trains and tracks to identify potential issues, such as brake wear or track defects. Rail operators can improve safety, reduce maintenance costs, and increase operational efficiency by addressing these issues before they lead to failures.

AI can also enhance CBM through automated fault diagnosis. Using AI, CBM systems can not only predict when a failure might occur but also diagnose the root cause of the problem. This can significantly speed up the maintenance process and reduce the time and cost required to get equipment back up and running.

In summary, AI can significantly enhance the effectiveness of Condition-Based Maintenance by providing more accurate predictions and diagnoses, optimizing maintenance schedules, reducing operational costs, and improving asset lifespan.

The terms "predictive maintenance" and "cognitive maintenance" can sometimes be confusing, as they are not always used consistently. Both involve using data analysis and predictive modeling to anticipate and prevent equipment failures, and both can potentially involve using artificial intelligence (AI). However, some distinctions can be made:

Predictive Maintenance: This typically involves using statistical methods, machine learning algorithms, and/or AI to analyze structured data (like sensor readings) to predict when equipment might fail. The goal is to perform maintenance just in time to prevent this. The predictive models used in this type of maintenance are often based on specific, predefined parameters.

Cognitive Maintenance: This is often considered a more advanced form of predictive maintenance. It uses AI and cognitive computing, which means the system is capable of "learning" from the data it processes and improving its predictive accuracy over time. Furthermore, cognitive maintenance systems can handle unstructured data, such as natural language text from maintenance logs or technician notes, and even images or sounds. This allows them to gain a more comprehensive understanding of equipment conditions and potential issues.

So, while AI can enhance predictive maintenance, not all AI-enhanced predictive maintenance would be considered "cognitive." The term "cognitive maintenance" is typically reserved for systems that can handle and learn from a wider variety of data types, continuously improve their performance, and perhaps even understand and interact with humans more naturally and intuitively. However, these distinctions can be somewhat blurry, and different sources may use the terms in slightly different ways.

Here is a table that summarizes the key differences between predictive maintenance and cognitive maintenance:

Feature	Predictive Maintenance	Cognitive Maintenance
Data used	Historical data and sensor data	Historical data, sensor data, and contextual information
Goal	To predict when an asset will fail	To predict when an asset will fail and to recommend how to improve the maintenance process
Accuracy	Can be very accurate, but can miss some patterns	More accurate than predictive maintenance, as it can identify patterns that predictive maintenance might miss
Cost	Relatively low	More expensive than predictive maintenance
Benefits	Can help to improve the reliability and efficiency of assets	Can help to improve the reliability and efficiency of assets even further than predictive maintenance

Cognitive Asset Maintenance (CAM) represents a significant leap forward in applying Artificial Intelligence (AI) to asset management. It encompasses a suite of technologies and methodologies that aim to improve asset maintenance procedures' efficiency, effectiveness, and predictability. The key to CAM lies in its ability to mimic human cognition – to understand, learn, reason, and even exhibit a level of intuition about the operational conditions of an asset.

Understanding Cognitive Maintenance

At its core, CAM leverages AI, machine learning, and advanced analytics to create an asset management system that can not only analyze and learn from vast amounts of data in real-time but also predict future performance and potential issues. This predictive capability is crucial, as it allows for proactive maintenance – a move away from the traditional reactive approach to asset management.

The CAM approach entails creating a digital twin of an asset, a virtual representation of the physical asset in the digital world. The digital twin is continually updated with real-time data from various sensors installed on the physical asset, allowing for ongoing monitoring and analysis of the asset's condition.

This cognitive approach to asset maintenance has the potential to revolutionize asset management, offering unprecedented levels of insight into asset performance and health. This, in turn, can lead to significant improvements in operational efficiency, cost savings, and asset longevity, fundamentally changing how organizations manage their physical assets.

In the context of a wastewater treatment utility, Cognitive Maintenance involves using AI and ML to analyze vast amounts of data collected from various sensors installed on equipment and infrastructure. This analysis enables the system to understand normal operational patterns, detect anomalies, predict potential failures, and suggest optimal maintenance schedules. This level of insight, coupled with the ability to act proactively, can significantly enhance the efficiency and reliability of wastewater treatment operations.

One key aspect of Cognitive Maintenance is the creation of a "digital twin" – a virtual replica of the physical asset in a digital space. This digital twin, constantly updated with real-time data, enables the AI to monitor, learn from, and predict the asset's performance and condition. This technology is precious in wastewater treatment facilities, where the early detection of equipment issues can prevent operational disruptions and potential environmental incidents.

The application of Cognitive Maintenance in wastewater treatment utilities is revolutionizing the sector. By enabling a shift from reactive to proactive and predictive maintenance, it can lead to significant operational improvements, cost savings, and enhanced environmental compliance. As such, understanding and implementing Cognitive Maintenance is a key step toward the future of wastewater treatment asset management.

Cognitive maintenance is an advanced form of predictive maintenance. Artificial Intelligence (AI) and Machine Learning (ML) are employed to analyze and learn from vast amounts of data collected from assets and use this knowledge to predict potential failures and optimize maintenance schedules. Here's how AI plays a crucial role in cognitive maintenance:

Advanced Anomaly Detection: AI, specifically machine learning, can identify patterns and trends within large data sets. For example, it can analyze data collected from IoT sensors on a piece of machinery to identify what 'normal' operation looks like. From this learned model of normality, the AI can detect anomalies or deviations, which could indicate an impending failure.

Example: An industrial machinery company implemented cognitive maintenance using AI for their production lines. The AI system was trained to understand normal machinery operations. It could subsequently detect anomalies indicative of potential mechanical issues, such as unusual vibrations or temperature fluctuations. This enabled MechPro to address the issues proactively and prevent unscheduled downtime.

Predictive Analytics: AI-based predictive analytics is the backbone of cognitive maintenance. The AI system detects anomalies and can predict when a failure is likely to happen based on the observed data trends and patterns.

Example: In a hydroelectric power plant, an AI system was set up to monitor the condition of turbines. The system could estimate the time to failure for different turbine components using predictive analytics, allowing the plant to schedule maintenance just in time to prevent failures, thus minimizing downtime.

Optimizing Maintenance Schedules: AI in cognitive maintenance can learn from historical maintenance data and equipment operational parameters to optimize maintenance schedules. It can predict when maintenance will be required based on the equipment's current condition, allowing organizations to transition from time-based to condition-based maintenance schedules.

Example: A city's public transportation system uses an AI system for the cognitive maintenance of its fleet of buses. The AI system used data from onboard sensors to predict when each bus would require maintenance, allowing the city to optimize its maintenance schedule, reducing unnecessary maintenance, and ensuring bus service reliability.

Self-learning Systems: One of the unique characteristics of cognitive maintenance is the ability of the AI system to learn from new data continuously. As the system is exposed to more data, it refines its models and predictions, improving its accuracy over time.

Example: A manufacturing plant used an AI-based cognitive maintenance system to monitor its robotic assembly lines. The system learned from each maintenance cycle, continually updating its model of normal machine operation and improving its predictive accuracy. This continuous learning process resulted in an ever-improving maintenance strategy, ensuring the plant's operations remained efficient and reliable.

In summary, AI plays a significant role in cognitive maintenance, offering advanced anomaly detection, predictive analytics, optimized maintenance schedules, and self-learning capabilities. These capabilities can improve asset reliability, reduce maintenance costs, and increase operational efficiency.

Cognitive maintenance is an advanced maintenance strategy involving artificial intelligence (AI) and machine learning (ML) to predict and prevent asset failures. It goes beyond the capabilities of predictive maintenance by not only predicting when an asset might fail but also understanding why it might fail and recommending specific actions to prevent it. Cognitive maintenance utilizes vast amounts of data from various sources, such as IoT sensors, maintenance logs, and operational data, to learn patterns and make intelligent predictions and recommendations.

Case Study: Large-scale Wastewater Treatment Facility

In a large-scale wastewater treatment facility, significant downtime and maintenance costs were being caused by the frequent failure of critical assets such as pumps and blowers. As a result, the facility decided to implement a cognitive maintenance strategy to tackle this issue.

The facility installed IoT sensors on their critical assets to continuously monitor parameters such as vibration, temperature, pressure, and flow rate. They also digitized their maintenance logs and operational data. This data was fed into an AI-powered cognitive maintenance system that had been trained to recognize patterns associated with different types of asset failure.

The cognitive maintenance system could identify patterns in the data that human operators had been unable to see. As a result, it was able to predict potential failures days or even weeks in advance, providing the maintenance team with ample time to take preventative action. Furthermore, the system understood why a failure was likely to occur and recommended specific maintenance actions to prevent the failure.

As a result of implementing cognitive maintenance, the wastewater treatment facility significantly reduced downtime and maintenance costs. The lifespan of their critical assets was extended, and the overall efficiency and reliability of their operations were improved.

In summary, Cognitive Maintenance represents the cutting edge of asset management strategies. By leveraging AI and ML technologies, it provides unprecedented predictive capabilities and actionable insights, enabling organizations to optimize their maintenance operations and significantly enhance the performance and lifespan of their assets.

Case Study: Cognitive Maintenance at a Manufacturing Company

A large Manufacturing Company, faced with unexpected machine downtime, took a step forward by adopting cognitive maintenance. This advanced predictive maintenance system relies on artificial intelligence (AI) and Internet of Things (IoT) technologies.

This state-of-the-art system was set up with IoT sensors affixed to many critical assets in their manufacturing line, such as high-speed presses, conveyor belts, and hydraulic units. These sensors gathered vast amounts of data, including temperature, vibration, humidity, noise, and pressure. The collected data was then transmitted to a cloud-based platform.

On this cloud platform, advanced machine learning algorithms were employed. These AI-driven algorithms were designed to understand the normal operational behavior of each machine, identify distinct patterns, and promptly detect anomalies that could indicate potential equipment failures.

Over time, as the AI processed more data and increased its knowledge about the machinery's performance, the system evolved and became progressively better at predicting when a machine was likely to require maintenance. Consequently, the Manufacturing Company was able to switch from a traditional time-based preventive maintenance schedule to a more efficient and cost-effective condition-based maintenance approach. This means maintenance was conducted only when the data indicated it was necessary, reducing unnecessary servicing and potentially extending the assets' lifespan.

Furthermore, with early detection of potential machine faults, many issues could be rectified before leading to a full-blown machine failure. The cost of preventive repairs often pales compared to the expense of replacing a failed machine, thus providing significant cost savings.

The cognitive maintenance system also improved the safety standards within the company by enabling the prediction of failures that could pose a risk to personnel. Lastly, the large volume of data collected and analyzed by the cognitive maintenance system provided a holistic view of the assets' health, enabling data-driven decision-making about asset management and replacement.

The Manufacturing Company significantly enhanced its asset maintenance strategy by implementing this cognitive maintenance system. It took a step towards becoming a more future-ready enterprise in a rapidly evolving manufacturing landscape.

Artificial Intelligence in Enhancing Inventory Management

How AI is transforming inventory management for asset maintenance

Artificial intelligence (AI) is transforming how inventory management is handled, especially in the context of asset maintenance. AI can analyze large amounts of data, make predictions, and automate tasks, leading to significant improvements in inventory management.

Here are some ways AI can enhance inventory management for asset maintenance:

- **Predictive analytics for inventory planning:** AI can analyze historical usage patterns, maintenance schedules, and other factors to predict future inventory needs. This can help maintenance teams ensure that the right parts are available when needed, reducing downtime and improving efficiency.
- **Automated reordering:** AI algorithms can monitor inventory levels in real-time and automatically reorder parts when they fall below a certain threshold. This can prevent stockouts and overstocking, optimizing inventory levels.
- **Optimized repair vs. replace decisions:** By analyzing data on part costs, failure rates, repair times, and other factors, AI can help maintenance teams make optimal decisions on whether to repair or replace a failing part. This can lead to cost savings and improved asset availability.
- **Vendor performance analysis:** AI can analyze vendor delivery times, quality, and costs data to help maintenance teams choose the best vendors and negotiate better terms. This can lead to better resource allocation and cost savings.
- **Anomaly detection:** AI can detect anomalies in inventory data, such as sudden increases in using a particular part, which could indicate an emerging maintenance issue. This can help maintenance teams take preventive measures to avoid costly downtime.
- **Enhanced forecasting:** AI can analyze patterns and trends in maintenance data to predict future needs and schedule preventive maintenance activities. This can lead to better resource allocation and cost savings.
- **Integration with IoT:** When coupled with Internet of Things (IoT) sensors on equipment, AI can predict when parts will fail and automatically order replacements, achieving truly predictive maintenance. This can help maintenance teams avoid costly downtime and improve asset availability.

Here are some specific examples of how AI is being used to enhance inventory management for asset maintenance:

- **GE Aviation:** GE Aviation uses AI to predict when aircraft components are likely to fail. This information is then used to ensure that the right spare parts are available when needed, which helps to reduce downtime and improve asset availability.
- **Siemens:** Siemens is using AI to optimize inventory levels for its wind turbines. This has helped the company to reduce inventory costs by 20% and improve customer satisfaction by 10%.

- **United Airlines:** United Airlines is using AI to forecast demand for spare parts. This information is then used to ensure that there are enough parts on hand to meet demand, which has helped the company to reduce stockouts by 50%.

By leveraging AI, companies can achieve a higher level of automation and intelligence in inventory management, leading to more efficient and effective maintenance operations.

AI in Demand Forecasting

Demand forecasting is the process of predicting future demand for a product or service. This is an essential part of inventory management, as it helps businesses ensure they have enough inventory to meet demand without overstocking.

Artificial Intelligence (AI) is revolutionizing demand forecasting by enhancing its accuracy and the speed at which predictions are made. AI, especially machine learning algorithms, can consider a vast range of factors, some of which might be overlooked in traditional statistical methods, and learn from patterns in the data to predict future demand. Here's a more detailed look at how AI, particularly machine learning, is used in demand forecasting.

Traditionally, demand forecasting has been based on historical data and statistical models. However, AI is now being used to improve demand forecasting accuracy. AI-powered demand forecasting models can use various data sources, including historical, weather, social media, and economic data. They can also use advanced algorithms to learn from historical data and make more accurate predictions.

One of the most common AI algorithms used for demand forecasting is **ARIMA** (Autoregressive Integrated Moving Average). ARIMA is a statistical model that uses past data to predict future values. It is a relatively simple model, but it can be very effective for forecasting demand for products or services with a relatively stable demand pattern.

Another AI algorithm that is often used for demand forecasting is **neural networks**. Neural networks are a type of machine learning algorithm that can learn from data and make predictions. They are more complex than ARIMA models but can often provide more accurate predictions.

In addition to ARIMA and neural networks, several other AI algorithms can be used for demand forecasting. These include **support vector machines**, **decision trees**, and **random forests**.

The choice of which AI algorithm to use for demand forecasting depends on several factors, including the type of data available, the complexity of the demand pattern, and the desired level of accuracy.

Supervised Learning for Demand Forecasting

Machine learning algorithms used for demand forecasting are typically supervised, which means they learn from labeled data - in this case, historical inventory data with corresponding demand levels. One popular supervised learning algorithm used for demand forecasting is the **Random Forest** algorithm.

Random Forest is an ensemble learning method that operates by constructing multiple decision trees during training and outputting the mean prediction of the individual trees for regression tasks or the class that is the mode of the classes for classification tasks.

Assuming X_train is your training features and y_train is your training target (historical demand), you could use the Random Forest algorithm in Python's scikit-learn library like this:

```python
from sklearn.ensemble import RandomForestRegressor

# Create a random forest Regressor
rf = RandomForestRegressor(n_estimators=100, random_state=42)

# Train the model
rf.fit(X_train, y_train)
```

You can then use the trained model to predict future demand (X_test would be your features for which you want to predict demand):

```python
# Make predictions
predictions = rf.predict(X_test)
```

Feature Engineering

Feature engineering is a crucial step in demand forecasting. Features could include past demand, seasonality (time of year, day of week, etc.), promotions, price changes, and more. For example, consider a time series of past demand. Feeding the series directly into a machine learning model might not be sufficient, especially if the demand is seasonal or has other temporal patterns. In this case, it might be helpful to create additional features such as:

- **Lagged features:** Demand from previous days (lag of 1 day, lag of 2 days, etc.)
- **Rolling window features:** Mean or sum of demand in a rolling window (e.g., rolling mean of the past 7 days)
- **Expanding window features:** Mean or sum of all past data up to the current point

Hyperparameters are parameters not learned from the data and must be set prior to training. For the Random Forest algorithm, hyperparameters include the number of trees in the forest (n_estimators) and the maximum depth of the trees (max_depth), among others.
Tuning these hyperparameters can improve the model's performance. This is usually done through a process like grid search or random search, where different combinations of hyperparameters are tested to find the combination that performs best.

Here's an example of how to perform hyperparameter tuning with GridSearchCV in scikit-learn:

```python
from sklearn.model_selection import GridSearchCV

# Define the parameter grid
param_grid = {
    'n_estimators': [50, 100, 200],
    'max_depth': [None, 10, 20, 30],
}

# Create a base model
rf = RandomForestRegressor(random_state=42)

# Instantiate the grid search model
grid_search = GridSearchCV(estimator=rf, param_grid=param_grid,
                           cv=3, n_jobs=-1, verbose=2)

# Fit the grid search to the data
grid_search.fit(X_train, y_train)

# Get the best parameters
best_params = grid_search.best_params_
```

With these concepts of supervised learning, feature engineering, and hyperparameter tuning, you can leverage AI to build highly accurate demand forecasting models to improve your inventory management process significantly.

Predictive Analytics for Stock Optimization

Predictive analytics uses data and statistical models to predict future events. This can be used to optimize stock levels by predicting future demand. Stock optimization is a key component of inventory management. It involves balancing the cost of inventory with the service level to customers. Too much stock can lead to high inventory holding costs, while too little can lead to lost sales and service level issues.

Several Machine learning algorithms can be used for predictive analytics, including **regression, classification**, and **clustering**. Regression algorithms can be used to predict continuous values, such as the number of units that will be sold in the next month. Classification algorithms can be used to predict categorical values, such as whether a customer will buy a product or not. Clustering algorithms can be used to group customers together based on their buying behavior.

Machine learning algorithms can be trained on historical sales, weather data, and social media data to learn the patterns that drive demand. Once they have learned these patterns, they can be used to predict future demand.

Another way AI can be used in predictive analytics for stock optimization is **through simulation** models. Simulation models can test different stocking strategies and see how they affect future demand. This can help businesses to optimize their stock levels and avoid overstocking or stockouts.

Once a machine learning algorithm has been trained on historical data, it can be used to predict future demand. This information can then be used to optimize stock levels by ensuring enough inventory to meet demand without overstocking.

Here's how you might implement a predictive analytics solution for stock optimization:

Understanding the Data

Typical datasets for this task would include historical sales data, possibly supplemented with additional information such as promotional activities, pricing changes, and external factors like holidays or economic indicators. The target variable is the number of products sold (or demanded) in a certain period, and the features would be the additional information mentioned.

```python
import pandas as pd

# Load data
data = pd.read_csv("sales_data.csv")

# Explore data
data.head()
```

In addition to the features present in the data, you may want to engineer additional features such as the lagged sales (sales in the previous weeks), rolling mean of sales, time since the last promotion, and others. These could help capture trends, seasonality, and the effect of promotions.

```python
# Create lagged feature
data['sales_lag_1'] = data['sales'].shift(1)

# Create rolling mean feature
data['sales_roll_mean_7'] = data['sales'].rolling(window=7).mean()
```

Model Development

A common model used for demand prediction is the XGBoost, a gradient boosting framework that can be used for both regression (predicting a continuous output) and classification (predicting a categorical output).

Let's use it to predict future sales:

```python
from xgboost import XGBRegressor
from sklearn.model_selection import train_test_split
from sklearn.metrics import mean_absolute_error

# Define features and target
X = data.drop(['sales'], axis=1)
y = data['sales']

# Split data into train and test sets
X_train, X_test, y_train, y_test = train_test_split(X, y, test_size=0.2,
random_state=42)

# Define model
model = XGBRegressor(n_estimators=1000, learning_rate=0.05)

# Fit model
model.fit(X_train, y_train,
          early_stopping_rounds=5,
          eval_set=[(X_test, y_test)],
          verbose=False)

# Make predictions
predictions = model.predict(X_test)

# Evaluate model
mae = mean_absolute_error(predictions, y_test)
```

Once you have a demand prediction, you can use it to set optimal stock levels. A simple way to do this would be to set the stock level to the predicted demand plus a safety stock to cover for forecast errors and demand variability.

The safety stock can be calculated using the following formula:

```
Safety Stock = Z-score * Standard Deviation of Lead Time * Demand Variability
```

Where:
- The Z-score is a factor of service level. For instance, a Z-score of 1.28 corresponds to a service level of 90%.
- The Standard Deviation of Lead Time is a measure of the variability in lead time.
- Demand Variability is the standard deviation of demand during the lead time.

This formula assumes that demand and lead time are normally distributed, which may not always be true. However, it can still serve as a good starting point.

Finally, the optimal stock level can be calculated as follows:

```
Optimal Stock Level = Predicted Demand + Safety Stock
```

This approach ensures you have enough stock to meet predicted demand while accounting for forecast errors and demand variability. It balances service level and inventory costs, leading to more efficient inventory management.

Obsolete and slow-moving inventory is inventory that is no longer needed, or that is not selling well. This type of inventory can drain a company's resources, as it takes up space and costs money to store.

AI can be used to manage obsolete and slow-moving inventory in several ways. One way is to use **machine learning** to identify items that are likely to become obsolete or slow-moving. This can be done by analyzing historical data, such as sales and inventory levels.

Once items likely to become obsolete or slow-moving have been identified, AI can **predict** when they will become obsolete or slow-moving. This information can then be used to take action, such as selling the items or donating them to charity.

AI can also **optimize** inventory levels for obsolete and slow-moving items. This can be done by analyzing historical data to determine the optimal number of items to keep in stock.

Here are some of the machine learning algorithms that can be used for managing obsolete and slow-moving inventory:
- **Linear regression** can be used to predict when items will become obsolete or slow-moving.
- **Decision trees** can be used to identify items that are likely to become obsolete or slow-moving.
- **Random forests** can be used to optimize inventory levels for obsolete and slow-moving items.

Here is an example of a code snippet that can be used to implement AI for managing obsolete and slow-moving inventory:

Understanding the Data

The typical data needed for this task would be sales history for each item, possibly enriched with additional features such as the item's age, price, category, and others. The target variable would be a binary flag indicating whether the item becomes slow-moving or obsolete in a certain period.

Here are some of the data that need to be collected to manage obsolete and slow-moving inventory:

- **Sales data:** This data can be used to track the sales of individual items over time.
- **Inventory levels:** This data can be used to track the number of items in stock.
- **Cost data:** This data can be used to track the cost of individual items.
- **Expiration dates:** This data can be used to track the expiration dates of individual items.

Once the data has been collected, it needs to be **cleaned** and **prepared** for analysis. This includes removing any errors or inconsistencies in the data, and transforming the data into a format that can be used by the AI algorithms.

Feature Engineering

The next step is to **select** the features that will be used to train the AI models. The features that are selected should be those that are most likely to be predictive of whether an item will become obsolete or slow-moving. You might want to create additional features that could be predictive of an item becoming slow-moving or obsolete. For example, you could create a feature that measures the trend in sales over time or a feature that captures the seasonality of the item's sales.

Model Development

Once the features have been selected, the AI models can be trained. The training process involves feeding the data to the models and allowing them to learn from the data.

Once the models have been trained, they can be used to make predictions about whether individual items will become obsolete or slow-moving. The predictions can then be used to take action, such as selling the items or donating them to charity. A suitable model for this task is a binary classification model that predicts whether an item will become slow-moving or obsolete. One popular choice for such tasks is the Random Forest Classifier.

Here is an example of the code snippet that can be used to implement AI for managing obsolete and slow-moving inventory:

```python
import pandas as pd
import numpy as np
from sklearn.linear_model import LinearRegression
from sklearn.tree import DecisionTreeClassifier
from sklearn.ensemble import RandomForestRegressor

# Load the data
data = pd.read_csv("inventory_data.csv")

# Create the independent variables
x = data[["historical_sales", "inventory_levels"]]

# Create the dependent variable
y = data["obsolete_or_slow_moving"]

# Create the models
linear_regression_model = LinearRegression()
decision_tree_model = DecisionTreeClassifier()
random_forest_model = RandomForestRegressor()

# Fit the models
linear_regression_model.fit(x, y)
decision_tree_model.fit(x, y)
random_forest_model.fit(x, y)

# Make predictions
linear_regression_predictions = linear_regression_model.predict(x)
decision_tree_predictions = decision_tree_model.predict(x)
random_forest_predictions = random_forest_model.predict(x)

# Print the predictions
print(linear_regression_predictions)
print(decision_tree_predictions)
print(random_forest_predictions)
```

A Comprehensive Guide to AI Implementation for Predictive Maintenance

Embarking on the journey to develop an in-house AI and Machine Learning-based predictive maintenance system is intricate but immensely fulfilling, particularly in our current data-centric era. The primary goal is to leverage machine learning (ML) and artificial intelligence (AI) to develop predictive models that utilize historical and real-time data from various sources, including Computerized Maintenance Management Systems (CMMS), Internet of Things (IoT) sensors, and SCADA systems. Implementing an in-house predictive maintenance system using AI and machine learning presents a myriad of advantages:

1. **Data Control**: Full control over data is a primary advantage of in-house development. It enables the organization to effectively manage its data privacy and security without relying on third-party vendors.
2. **Customization**: In-house development allows for greater customization and flexibility. Every organization is unique, with specific needs and workflows. An in-house system can be tailored to fit these exact requirements.
3. **Cost Efficiency**: While the upfront cost might be higher, maintaining and updating an in-house system can be more cost-efficient in the long run than recurring vendor fees.
4. **Integration**: Proprietary systems are often better equipped to integrate with existing processes and systems. This is especially important when dealing with complex IoT sensors and SCADA systems.
5. **Skill Development**: Building an in-house solution can strengthen your team's technical capabilities, fostering a culture of innovation and learning.
6. **Direct Oversight**: With in-house development, the organization directly oversees the entire process, from data collection and preprocessing to model development and validation.
7. **Continuous Improvement**: An in-house system can adapt to changes more swiftly, enabling the company to continuously improve the predictive maintenance models as more data accumulates and business needs evolve.

Implementing an artificial intelligence (AI) system for predictive maintenance involves several steps. The objectives of the predictive maintenance system must be clearly defined. What equipment will it monitor, and what types of failures will it predict? What data is currently being collected, and what additional data might be needed?

The predictive maintenance journey begins by defining the system's objectives. This step is crucial to aligning the system with your organization's operational context, goals, and constraints. A wastewater treatment plant, for example, might aim to minimize downtime, optimize maintenance routines, extend equipment lifespan, or meet stringent regulatory requirements.

Once the overall objectives are defined, the next step is identifying and prioritizing the equipment for monitoring. In a wastewater treatment plant, this could include a range of equipment from pumping systems and aeration equipment to sedimentation tanks and chemical dosing systems. The selection criteria should consider factors such as the criticality of the equipment, the cost of downtime, and the frequency and impact of failure.

After identifying the equipment, we define the specific objectives for monitoring each piece. For instance, monitoring a pump might be to prevent catastrophic failures that could cause significant downtime and environmental impact. On the other hand, monitoring an aeration system might be focused on optimizing energy use and meeting regulatory standards for wastewater treatment.

The type and amount of data required for predictive maintenance depend on the defined objectives. We need data on parameters like operating pressure, vibration levels, and power consumption for pumping systems. Data on dissolved oxygen levels, airflow rates, and energy usage might be required for aeration systems.

These initial steps lay the groundwork for a successful predictive maintenance program, setting clear goals, identifying key equipment, and determining data requirements. In the following sections, we will delve deeper into the process, covering steps such as data collection, data preprocessing, feature selection, AI model selection, model implementation and training, and translating AI predictions into actionable insights. Each step will be contextualized within the wastewater treatment plant scenario, providing a comprehensive guide for AI implementation in predictive maintenance.

The first step is to define the system's objectives when setting up a predictive maintenance system. These objectives should encapsulate your goals and consider the specific context and constraints of your operation.

Let's take the example of a wastewater treatment plant. This type of facility presents unique challenges and requirements, such as maintaining continuous operation, meeting strict environmental regulations, and managing a range of complex, heavy-duty equipment.

Given these characteristics, the objectives of a predictive maintenance system for a wastewater treatment plant could be as follows:

1. **Maintain Continuous Operation:** Wastewater treatment is a 24/7 operation. Any equipment failure can cause significant disruption and pose environmental risks. Therefore, one primary objective of the predictive maintenance system could be to prevent unexpected breakdowns and maintain continuous operation.

2. **Meet Environmental Regulations:** Wastewater treatment plants operate under stringent environmental regulations. Equipment malfunctions can lead to non-compliance issues, such as untreated or improperly treated water being released. Thus, another objective could be to use predictive maintenance to ensure all equipment operates correctly and within regulatory standards.

3. **Optimize Maintenance Resources:** Wastewater treatment plants often have a wide variety of equipment, from pumps to sedimentation tanks. Each piece of equipment may require different maintenance strategies. Therefore, an objective could be to use predictive maintenance to optimize maintenance resources, ensuring that maintenance activities are targeted and efficient.

4. **Extend Equipment Lifespan:** Replacing equipment in a wastewater treatment plant can be costly and disruptive. Therefore, another objective might be to use predictive maintenance to detect early signs of wear and tear, allowing for proactive maintenance that can extend the equipment's lifespan.

5. **Improve Worker Safety:** Heavy machinery and hazardous conditions can pose safety risks to workers in a wastewater treatment plant. Thus, improving worker safety could be another objective of the predictive maintenance system, with early detection of potential equipment failures reducing the risk of accidents.

In a predictive maintenance program, it's crucial to identify and prioritize the equipment for monitoring. Not all equipment may be suitable for or require predictive maintenance. The decision to monitor specific equipment should be based on factors such as criticality, frequency of failure, cost of downtime, and so on.

Returning to the example of a wastewater treatment plant, several types of equipment may warrant consideration for predictive maintenance:

1. **Pumping Systems:** These are critical for the functioning of a wastewater treatment plant. A pump failure can halt the treatment process and possibly lead to environmental compliance issues. Given their importance, the criticality of pumps makes them prime candidates for monitoring.

2. **Aeration Equipment:** Aeration equipment supplies oxygen to the treatment process and is another critical component. If this equipment fails, it can severely disrupt the biological processes central to wastewater treatment.

3. **Sedimentation Tanks:** These tanks are used to separate solids from the water. While they may have a lower failure frequency than pumps or aeration systems, their downtime can still significantly impact the plant's operation.

To prioritize equipment for monitoring, a few criteria might be considered:

- **Criticality:** Some equipment is essential to the plant's operation, and its failure could cause significant disruption. For instance, pumps and aeration equipment are usually of high criticality in a wastewater treatment plant.

- **Failure Frequency:** Equipment that fails frequently or requires regular maintenance can be an excellent candidate for predictive maintenance. The data from these frequent failures can be useful in predicting future problems.

- **Cost of Downtime:** Equipment whose failure would result in significant downtime or costly repairs should be prioritized for monitoring.

- **Safety Considerations:** If the failure of a piece of equipment could potentially lead to a safety issue, it should be high on the monitoring priority list.

By identifying and prioritizing equipment based on these criteria, a wastewater treatment plant can focus its predictive maintenance efforts where it will provide the most benefit. This approach ensures that the plant achieves its operational objectives effectively and efficiently.

Different pieces of equipment in a wastewater treatment plant have distinct roles and are subject to unique operational conditions. As a result, the reasons for monitoring each piece of equipment can vary. Objectives for monitoring may range from preventing catastrophic failures and optimizing routine maintenance to meeting regulatory requirements. Clearly defining these objectives for each piece of equipment is crucial in implementing an effective AI-based predictive maintenance system.

Preventing Catastrophic Failures
Some equipment in a wastewater treatment plant, such as pumps, blowers, and electrical panels, are critical to the operation of the entire plant. A failure in any of these pieces of equipment can cause a shutdown of the entire plant, resulting in significant operational and financial impacts. The primary objective for monitoring such equipment is to prevent such catastrophic failures.
For example, sludge pumps, essential for removing solid waste, might be monitored to predict bearing failures or impeller blockages. Predictive maintenance can help identify the early signs of these issues, allowing for preventative action before a catastrophic failure occurs.

Optimizing Routine Maintenance
Other equipment, such as filters, mixers, or UV disinfection systems, may not lead to catastrophic failures if they malfunction. However, their efficient operation is crucial for the plant's overall efficiency. The objective of monitoring these types of equipment could be to optimize routine maintenance schedules, reduce unnecessary maintenance activities, and extend the equipment's operational life.
For instance, monitoring the UV intensity of a UV disinfection system can provide insights into when the UV lamps are likely to need replacement. Instead of following a rigid replacement schedule, predictive maintenance can enable more flexible, condition-based maintenance, potentially reducing costs and minimizing downtime.

Meeting Regulatory Requirements
Some equipment may need to be monitored to ensure compliance with environmental regulations. For example, equipment involved in the treatment processes that directly impact the quality of the effluent, such as biological reactors or chemical dosing systems, could be monitored to ensure that the treated wastewater meets regulatory standards.
In the case of biological reactors, predictive maintenance can help optimize the operation of the aeration system, ensuring that the biological treatment process works effectively to remove pollutants and that the effluent quality remains within the allowed limits.

By defining clear objectives for monitoring each piece of equipment, wastewater treatment plants can tailor their predictive maintenance strategies to their specific needs and constraints, maximizing the benefits of their AI-based predictive maintenance systems.

When implementing a predictive maintenance strategy in a wastewater treatment plant, the type and amount of data required will largely depend on each equipment piece's specific objectives. Here are some examples of the types of data you might need to collect:

1. **Pumping Systems:** The data required for predictive maintenance of pumping systems might include operating pressure, flow rates, vibration levels, and power consumption. These parameters can provide insights into the pump's performance and potential issues. The frequency of data collection might be every few seconds or minutes, depending on the pump's criticality and the failure modes you're trying to predict.

2. **Aeration Equipment:** For aeration systems, you might need to collect data such as dissolved oxygen levels, airflow rates, and energy usage. This data can help identify inefficiencies or problems that might lead to failure if not addressed. You might collect this data every few minutes or hours, depending on the specific equipment and objectives.

3. **Sedimentation Tanks:** For sedimentation tanks, you might collect data on parameters such as inflow and outflow rates, sludge levels, and water clarity. This data can help ensure the tanks operate efficiently and meet regulatory requirements. Depending on the regulatory standards and operational practices, the data might be collected at regular intervals, such as every hour or day.

Data collection is the backbone of any AI-based predictive maintenance system. The quality and range of the data collected will directly impact the effectiveness of the AI models. This data can be gathered from various sources within a wastewater treatment plant, including IoT sensors and Supervisory Control and Data Acquisition (SCADA) systems.

Firstly, determine the necessary data collection mechanisms. IoT sensors could be a viable option for capturing real-time data about the state of equipment. For instance, you might use vibration and temperature sensors on pump systems to monitor their health. In the context of aeration systems, dissolved oxygen sensors and flow meters could be deployed to monitor performance and efficiency.

When choosing IoT sensors, consider the type of data required for your specific predictive maintenance goals. The sensors' precision, durability, and reliability are critical to ensuring the long-term collection of high-quality data.

In addition to IoT sensors, you might already have a SCADA system in place, which can be a rich source of data for predictive maintenance. SCADA systems are typically used in industrial environments like wastewater treatment plants for process control and to collect a wealth of operational data. This data can be leveraged for your predictive maintenance objectives.

Once your data collection mechanisms are in place, establish a system to aggregate and store this data. This could be a dedicated database or a cloud-based storage solution. Ensure that the data aggregation system can handle the volume and frequency of data being collected and incorporates data validation to identify and flag potential errors.

Lastly, have a contingency plan for dealing with potential data quality issues. Regular data audits, anomaly detection algorithms, and clear procedures for addressing data quality problems can help ensure your AI models have the best possible data to work with.

After collecting data from the various equipment and systems within a wastewater treatment plant, it's not uncommon to find that this raw data is messy and requires some cleaning and preprocessing before it can be used for AI predictive maintenance algorithms. This is a crucial step in the process, as data quality directly affects the accuracy and reliability of the AI models. Data preprocessing and cleaning can involve several tasks, such as:

1. **Dealing with missing values:** Data collected from IoT sensors and SCADA systems may sometimes have gaps due to network glitches, sensor malfunctions, or human errors. For example, a sensor on a pump might fail to record data due to a temporary power outage. When these missing values occur, several strategies can be employed to address them, such as imputation, where missing values are replaced with a calculated value (like the mean, median, or mode), or using predictive models to estimate the missing values based on other related data.

2. **Handling outliers:** Outliers are data points that deviate significantly from the norm. In a wastewater treatment plant, these could be unusually high readings from a pH sensor, perhaps due to a calibration issue or an unexpected spike in flow rate caused by a sudden equipment failure. Outliers can skew the predictive model and need to be identified and handled appropriately. This could mean discarding these values if they're deemed errors or analyzing them separately if they represent significant events.

3. **Reducing noise:** Noise in the data can come from sensor errors, electronic interference, or random environmental fluctuations. For example, a temperature sensor near a heat source might record fluctuating values. Techniques such as smoothing (moving averages, exponential smoothing) or signal processing techniques (Fourier transform, wavelet transform) can reduce noise and make the data more suitable for AI algorithms.

4. **Normalization or Scaling:** AI models often perform better when the input data is on a similar scale. Monitor both temperatures (in degrees) and flow rate (in thousands of gallons per minute). Those numbers will be on very different scales. Techniques such as min-max scaling or standardization can bring different types of data onto a comparable scale.

5. **Feature Engineering:** This involves transforming the raw data into a more suitable format for the AI models. For example, suppose you are trying to predict equipment failures. In that case, you might create a feature that represents the change in vibration level over the past hour or the average temperature over the past day.

This step involves assigning labels to the data. In the case of predictive maintenance, the labels are typically either "normal" or "failure." The data labeling process can be time-consuming and labor-intensive. However, it is crucial to ensure that the data is labeled accurately. This is because inaccurate data can lead to inaccurate predictions.

- **Vibration data:** Vibration data can be used to identify potential mechanical problems in a pump. For example, increased vibration can be a sign of bearing wear or misalignment.
- **Temperature data:** Temperature data can also identify potential mechanical problems in a pump. For example, increased temperature can be a sign of friction or overheating.
- **Pressure data:** Pressure data can also identify potential mechanical problems in a pump. For example, decreased pressure can indicate a leak or blockage.
- **Flow rate data:** Flow rate data can also identify potential mechanical problems in a pump. For example, the decreased flow rate can be a sign of blockage or impeller damage.
- **Power consumption data:** Power consumption data can also be used to identify potential mechanical problems in a pump. For example, increased power consumption can indicate friction or overheating.

By labeling data for these features, you can create a dataset that can be used to train a machine-learning model to predict mechanical failure in a pump.
Here are some additional tips for data labeling:

- **Use a clear and consistent labeling process.** The labeling process should be clear and consistent. This will help to ensure that the data is labeled accurately.
- **Use a quality assurance process.** A quality assurance process should be used to check the accuracy of the labeled data. This can be done by checking the labels against a known dataset.
- **Label the data promptly.** The data should be labeled as soon as possible after it is collected. This will help to ensure that the data is fresh and accurate.

Once the data has been cleaned and preprocessed, the next step in implementing AI for predictive maintenance within a wastewater treatment plant is feature engineering and selection. This process involves transforming the cleaned data into features (input variables) that are most relevant for the predictive model and selecting the most influential ones to optimize the performance of the AI algorithm.

Feature Engineering

Feature engineering is the process of creating new features or modifying existing ones to better represent the underlying patterns in the data relevant to the predictive task. These engineered features are often derived from the raw data but provide more insightful information to the AI models. The aim is to translate domain knowledge into variables that an AI algorithm can leverage.

Features could be engineered in a wastewater treatment plant from data collected from various sources. For instance:

1. **Time-based features:** If you have time-series data from sensors monitoring equipment, you might create features representing the change in readings over specific periods, such as hourly or daily averages, the rate of change over time, or the time since the last maintenance.
2. **Operational features:** These could be derived from the operational parameters of the equipment. For example, you could calculate the ratio of actual to optimal performance for a pump or the deviation of temperature or pressure from their set points.
3. **Composite features:** These are created by combining two or more raw features. For instance, if you're monitoring a motor's vibration and temperature, you might create a feature that represents the interaction between these two variables.

Feature Selection

Once you have created your features, not all may be useful for your AI model. Some might be redundant, or too closely correlated with each other, or simply not relevant to the predictive task. This is where feature selection comes in.

Feature selection is the process of selecting the most informative features for use in model construction. The aim is to reduce the dimensionality of the data, improve model performance, and decrease the computational cost of training the model. Techniques for feature selection can range from simple correlation analysis and mutual information to more complex methods like recursive feature elimination or using machine learning models like Lasso or Random Forests for their feature importance capabilities.

For instance, in a wastewater treatment plant, you might find that specific sensor readings are highly correlated and effectively provide the same information so that you can remove some of them from your dataset. Alternatively, you may identify that some features, like the time since the last maintenance, are particularly influential in predicting equipment failures and should be included in your model.

After preprocessing the data and conducting feature engineering and selection, the next step in implementing AI for predictive maintenance in a wastewater treatment plant is model selection and development. This step involves choosing the most suitable AI model for the predictive task, developing it using the selected features, and tuning it for optimal performance.

Model Selection
Choosing the right model depends on several factors, including the nature of your data, the problem you're trying to solve, the need for interpretability, and computational resources.
1. **Regression Models:** For predicting continuous outcomes, such as the remaining useful life of a pump or the time to failure, regression models such as Linear Regression, Support Vector Regression, or Random Forest Regression could be used.
2. **Classification Models:** If the aim is to predict whether a failure will occur within a specific time frame, classification models such as Logistic Regression, Decision Trees, or Gradient Boosting Machines might be more appropriate.
3. **Time-Series Models:** For data with strong temporal patterns, time-series models like ARIMA, State Space Models, or Recurrent Neural Networks could be considered.
4. **Deep Learning Models:** If you have a large amount of high-dimensional data, deep learning models like Convolutional Neural Networks (for image data from visual inspections) or Long Short-Term Memory Networks (for time-series data from sensors) could be employed.

Model Development
The next step is model development once the appropriate model has been selected. This involves training the model on a subset of your data (the training set), tuning its parameters to improve its performance, and testing the model on another subset of your data (the testing set) that it hasn't seen before.
In the context of a wastewater treatment plant, suppose you're using a Random Forest Classifier to predict whether a pump will fail in the next 30 days. You would train the model on historical sensor data and maintenance records, tune parameters such as the number of trees in the forest and the maximum depth of the trees, and then test the model on recent data to evaluate its predictive accuracy.

Model Complexity vs. Interpretability
An important aspect to consider when selecting and developing your model is the trade-off between model complexity and interpretability. Complex models like deep learning networks can capture intricate patterns in the data and often provide superior predictive performance. However, they are often called "black boxes" because their predictions are not easily interpretable.

On the other hand, simpler models like linear regression or decision trees may not perform as well on complex tasks. However, they offer more interpretability, which can be crucial for understanding the reasons behind a prediction and for gaining the trust of decision-makers and operators at the wastewater treatment plant.

After feature engineering and model selection, the next crucial steps are model training and validation. These steps ensure that your chosen AI model can effectively learn from the dataset and accurately predict equipment failures in a wastewater treatment plant.

Model Training

Model training involves feeding your selected model with the feature-engineered dataset, allowing it to learn patterns to predict the target variable. For example, if you are using a Support Vector Machine (SVM) model to predict the failure of a pump based on features like vibration level, temperature, and pressure, you would use historical sensor data to train the model. The model would analyze this data and establish relationships between the sensor readings and pump failures.

Model Validation

Model validation is a critical step that involves testing the trained model's predictive capabilities on a separate set of data, often referred to as the validation or testing set. This set should be different from the one used for training to ensure that the model's performance is evaluated on data it hasn't seen before, providing a more realistic measure of its predictive power.

There are several methodologies for model validation:
1. **Holdout Method:** In this method, the dataset is split into two parts: a training set and a testing set. The model is trained on the training set and then tested on the testing set.
2. **Cross-Validation:** This is a more robust method for dividing the dataset into 'k' subsets. The model is trained on k-1 subsets and tested on the remaining subset. This process is repeated k times with a different subset used as the testing set. The most common form of this is k-fold cross-validation, where k is often set to 5 or 10.
3. **Time-Series Cross-Validation:** Given that sensor data in a wastewater treatment plant is usually a time-series, this method can be particularly useful. It involves creating multiple training/test splits so that every data point is included in a test set exactly once.

The choice of validation method depends on factors such as the size and nature of your data. The performance of the model is then evaluated using appropriate metrics. For regression tasks, metrics like Mean Absolute Error (MAE) or Root Mean Squared Error (RMSE) could be used. For classification tasks, metrics such as accuracy, precision, recall, and the F1 score are often used.

For example, suppose your model predicts whether a wastewater pump will fail next week. In that case, you might use accuracy (the percentage of correct predictions) as a performance metric. However, suppose false negatives (predicting no failure when a failure actually occurs) are particularly costly. In that case, you might prioritize recall (the percentage of actual failures the model correctly predicts).

Once the AI model for predictive maintenance has been trained and validated, the next step is its implementation into the existing workflow of the wastewater treatment plant. This step ensures that the predictive insights provided by the model can be acted upon efficiently, leading to improved maintenance processes and operational efficiency.

Developing a User Interface
A user-friendly interface can make it easier for operators and maintenance personnel to interact with the predictive maintenance system. The interface could display real-time sensor readings, predicted equipment failure probabilities, and recommended maintenance actions.

For example, the interface could show a dashboard with a list of equipment, their current status, and any maintenance alerts. It could also provide visualizations of sensor data over time and the corresponding predicted failure probabilities. This would enable the staff to understand the state of the equipment quickly and make informed maintenance decisions.

Setting Up Alert Systems
An effective alert system is crucial for predictive maintenance. The AI model might predict a high likelihood of failure for a particular piece of equipment. However, suppose this information is not conveyed promptly and effectively to the relevant personnel. In that case, the benefits of predictive maintenance might be lost.

Alert systems could be emails, text messages, or notifications on the user interface. They should provide enough information to allow the recipient to understand the nature of the predicted issue and take appropriate action.

For example, an alert could be set up to notify the maintenance team when the failure probability for a critical pump exceeds a certain threshold. The alert could include details about the pump (e.g., location, model, last maintenance date), the predicted failure probability, and the key sensor readings contributing to the prediction.

Integration with Other IT Systems
Ideally, the predictive maintenance system should be integrated with other IT systems used in the wastewater treatment plant. This could include the plant's Computerized Maintenance Management System (CMMS), Supervisory Control and Data Acquisition (SCADA) system, or Enterprise Resource Planning (ERP) system.

For instance, when a maintenance alert is triggered, a work order could be automatically created in the CMMS, including all relevant details from the alert. This would streamline the maintenance process and promptly create work orders.

The predictive maintenance system could also be linked to the plant's inventory management system. This would enable the system to check the availability of necessary spare parts and include this information in the maintenance alerts or work orders.

The successful implementation of an AI-driven predictive maintenance system hinges on the ability to translate AI predictions into tangible maintenance tasks. This translation is crucial for the daily operations of a wastewater treatment plant. It can directly impact the efficiency and reliability of the plant's operations.

Defining Thresholds for Triggering Maintenance Activities

One of the first steps in this process is to define the thresholds that will trigger maintenance activities. These thresholds are typically based on the probability of equipment failure as predicted by the AI model. For example, if the model predicts a 70% likelihood of failure within the next month for a critical pump, this could trigger a preventive maintenance task. Thresholds can be tailored to the plant's specific needs and risk tolerance. For high-risk, critical equipment, you might set a lower threshold to ensure prompt action and minimize the risk of failure. For less critical equipment, a higher threshold might be acceptable.

Assigning Responsibilities

Once a maintenance task is triggered, it's essential to have a straightforward procedure in place for assigning responsibilities. The nature of the task, the skills required, and the task's urgency will typically determine who is responsible for carrying it out.

For instance, the AI system could flag a potential issue with the effluent quality due to predicted anomalies in the secondary clarifier operation. Upon receiving this alert, the plant manager could assign a team of technicians to inspect the clarifier, check the relevant sensors and controls, and perform any necessary maintenance tasks.

Establishing Maintenance Procedures

Alongside assigning responsibilities, specific procedures should be established for maintenance tasks. These procedures should provide clear, step-by-step instructions, ensuring maintenance tasks are performed consistently and effectively.

For example, suppose the AI model predicts an impending failure in a sludge pump due to excessive vibration levels. In that case, the maintenance procedure might involve checking the pump alignment, inspecting the bearings, and replacing worn parts. A well-defined procedure ensures that the AI prediction is addressed systematically, enhancing the effectiveness of the preventive maintenance task.

Monitoring and Feedback

After maintenance tasks are completed, it's important to monitor the performance of the equipment to verify that the issue has been resolved. This feedback loop can help to refine the AI model's predictions over time and improve the overall effectiveness of the predictive maintenance system.

By translating AI predictions into actionable maintenance tasks, wastewater treatment plants can proactively address equipment issues, reduce unplanned downtime, and enhance overall operational efficiency."

The implementation of an AI-based predictive maintenance system is not a one-and-done process. The system's effectiveness needs to be monitored continuously, and adjustments should be made based on the results of this monitoring. This continuous improvement process is essential for optimizing the system's performance and maintaining the operational efficiency of the wastewater treatment plant.

Monitoring the AI System's Performance
Regularly monitoring the AI system's performance is critical to ensure its accuracy and effectiveness. This can involve tracking metrics such as prediction accuracy, false alarm rates, and cost savings achieved through preventive maintenance.
For example, suppose the system predicts a potential failure in the aeration process due to abnormal oxygen levels. In that case, the actual condition of the aeration tanks and blowers should be checked. If the predicted failure doesn't occur as frequently as the AI model indicates, this could indicate a high false alarm rate, suggesting that the model might need to be retrained or refined.

Retraining or Tuning the AI Model
The AI model might need to be retrained or tuned periodically based on the changes in the plant's operation conditions or the equipment's performance characteristics.
For instance, if a new type of impeller is installed in the sludge pumps, this could affect the vibration patterns that the model uses to predict pump failure. In such a case, the model may need to be retrained with updated data that reflects these changes.

Updating the Feature Set
The features or variables the AI model uses may also need to be updated over time. New sensors might be installed, new types of equipment might be added, or you might discover that certain variables are more predictive of equipment failure than previously thought.
For example, if the addition of a new sensor provides data on the temperature of the biological reactor, this new feature could be incorporated into the AI model to improve its predictions for the biological treatment process.

Revising the Objectives
As the wastewater treatment plant evolves and the operational context changes, the objectives of the predictive maintenance system might also need to be revised. New regulatory requirements, changes in the plant's capacity, or shifts in maintenance strategies could all necessitate a reevaluation of the system's objectives.
For example, if the plant expands its capacity and adds new equipment, the predictive maintenance system may need to be expanded to cover these new assets. Similarly, suppose a

new regulation requires more frequent monitoring of specific parameters. In that case, the predictive maintenance system may need to be adjusted to meet these requirements.

From Theory to Practice: Implementing a Predictive Maintenance System for Pump Mechanical Failure Prediction

Having walked through the comprehensive steps of implementing AI for predictive maintenance, it's time to shift our focus from theory to practice. As we venture into the practical realm, we will narrow our focus on a specific piece of equipment – a pump. The objective is to predict pump mechanical failure using vibration analysis.

In the context of a pump in a wastewater treatment facility, a predictive maintenance system should be equipped to predict a variety of potential failures, including:

1. **Mechanical Failures:** These are often the most common type of failure in pumps. They can include bearing failures, seal failures, impeller damage, and pump casing damage. Predictive maintenance systems often use vibration analysis, acoustic monitoring, and thermal imaging to detect these problems early.
2. **Operational Failures:** These issues arise from how the pump is being operated. For example, the pump may be running too fast or too slow, cavitating, or operating under conditions it was not designed for. Predictive maintenance systems can identify these issues with process parameters such as flow rate, pressure, temperature, and pump speed.
3. **Electrical Failures:** Pumps are often driven by electric motors, which can also fail. Electrical failures can include issues like winding failures, insulation breakdown, and problems with the power supply. Predictive maintenance systems can use techniques like motor current analysis and power quality monitoring to detect these problems.
4. **Corrosion and Wear:** Pump parts may corrode or wear out over time. This can lead to failures if not addressed. Predictive maintenance systems can use techniques like oil analysis and ultrasound to detect signs of corrosion and wear.
5. **System Failures:** These are issues that arise from the larger system that the pump is part of. For example, there may be issues with the piping system, the control system, or the power supply. Predictive maintenance systems can use data from other parts of the system to help predict these failures.

Each type of failure has its unique indicators and requires different predictive strategies. In our case, we will focus on mechanical failure, which can often be predicted through vibration analysis.

Vibration analysis is a powerful tool for predicting mechanical failures because it can detect abnormalities such as misalignment, imbalance, and bearing faults. The data for vibration analysis includes measurements of vibration displacement, velocity, and acceleration, often in three axes (radial, axial, and tangential).

To implement our predictive maintenance system, we will use Python, a popular programming language for data analysis and machine learning, alongside various libraries, including pandas for data manipulation, NumPy for numerical computations, and scikit-learn for machine learning. The code provided in this chapter will be a simplified illustration of the process.

Let's dive into the details:

Step 1: Data Collection and Storage

The first step in building a predictive maintenance system is to collect and organize relevant data. In our case, we are focusing on a centrifugal pump to predict mechanical failure using vibration analysis.

The following table shows the Pump's measurement data:

Measurement Data	Unit of Measurement
Pump Vibration Amplitude	Meters (m), Millimeters (mm), Inches (in), or G's (gravitational units)
Pump Vibration Frequency	Hertz (Hz) or Cycles per Second (cps)
Pump Vibration Displacement	Meters (m), Millimeters (mm), or Inches (in)
Pump Vibration Velocity	Meters per Second (m/s), Millimeters per Second (mm/s), or Inches per Second (in/s)
Pump Vibration Acceleration	Meters per Second Squared (m/s^2), G's (gravitational units), or Inches per Second Squared (in/s^2)
Pump Harmonics	No specific unit, it is a multiple of a fundamental frequency
Pump Resonance Frequency	Hertz (Hz) or Cycles per Second (cps)

We have a sensor installed for this pump that collects vibration data at regular interval every second. This sensor measures the vibration displacement, velocity, and acceleration in three axes. Over time, this data forms our raw dataset.

Note: The data should be collected at a frequency appropriate for the application. For example, data may need to be collected more frequently for pumps operating in harsh environments or critical to a plant's operation.

Start by retrieving vibration data from your local SQL server and importing it to Google Cloud Storage. This is your raw data that will be used to predict mechanical failures in the centrifugal pump.

We'll use the google-cloud-bigquery library to interface with BigQuery, Google's fully managed, petabyte-scale data warehouse that allows super-fast SQL queries using the processing power of Google's infrastructure.

First, make sure you have the necessary libraries installed:

1. **Install the necessary Python libraries and Setup your google cloud service account to generate a JSON key file**

Python

```
pip install pyodbc pandas google-cloud-storage schedule
```

2. **Write the Python script that fetches data from SQL Server and uploads it to Google Cloud Storage.** Here's a skeleton of what that script might look like:

Note: You need to replace 'Your_Server_Name', 'Your_Database_Name', 'Your_User_ID', 'Your_Password', 'path_to_your_service_account_file.json' and 'your_bucket_name' with your actual values.

The script fetches data from your SQL Server, saves it to a CSV file, and then uploads the file to Google Cloud Storage. This is scheduled to run every day at a specific time (in the example, it's set to run every day at 01:00). The schedule can be adjusted to fit your specific needs.

This could involve removing outliers, handling missing values, and normalizing the data if necessary. Once the data is cleaned, you'll need to label it. Supervised learning model needs examples of normal operations and mechanical failures. You will have to manually label these instances in your dataset to provide 'ground truth' for your model. This process can be achieved using Google Cloud Data Labeling Service.

Install bigquery libraries if not already installed

```
pip install pandas google-cloud-bigquery google-cloud-storage pyodbc
```

Cleaning and Preprocessing

First, you need to remove outliers and handle missing values. Outliers might include any vibration readings that are unrealistically high or low. You might replace missing values with the median or mean values of the respective columns. You may also want to normalize the data if the scales of the different features vary widely.

```python
from google.cloud import bigquery
import pandas as pd

# Instantiate a BigQuery client
client = bigquery.Client()

# Define your BigQuery dataset and table
dataset_name = 'your_dataset_name'
table_name = 'PumpData'

# Fetch data into a pandas DataFrame
query = f'SELECT * FROM `{client.project}.{dataset_name}.{table_name}`'
df = client.query(query).to_dataframe()

# Handle missing values by replacing them with the mean of the column
df.fillna(df.mean(), inplace=True)

# Outlier removal using the Interquartile Range (IQR)
Q1 = df.quantile(0.25)
Q3 = df.quantile(0.75)
IQR = Q3 - Q1

df = df[~((df < (Q1 - 1.5 * IQR)) | (df > (Q3 + 1.5 * IQR))).any(axis=1)]

# Normalization of the data using Min-Max scaling
df = (df - df.min()) / (df.max() - df.min())
```

```python
        blob.upload_from_filename(file_path)

    print("File uploaded to {}.".format(bucket_name))


# Schedule the job every day at a specific time (e.g., 01:00)
schedule.every().day.at("01:00").do(job)


while True:
    schedule.run_pending()
    time.sleep(1)
```

Labeling

Supervised learning models require labeled data. We need to create a label for our data to indicate whether it's a normal operation or a mechanical failure.

Let's assume we have a domain expert who determined that any vibration above a certain threshold for a given frequency band indicates a mechanical failure. For simplicity, we'll assume that this threshold is a constant. However, in practice, it might vary depending on the specific characteristics of the pump.

```python
# Create a new column 'label' to store the labels
df['label'] = 0  # 0 indicates normal operation

# Define the threshold for vibration indicating a failure
failure_threshold = 0.8  # this is a hypothetical value and should be adjusted
according to your domain knowledge

# Label the data: if vibration is above the threshold, label it as a failure
(1)
df.loc[df['vibration'] > failure_threshold, 'label'] = 1
```

Now your data is cleaned, normalized, and labeled. The final DataFrame df can be used for further steps like feature engineering and model training.

Please adjust the code according to your data and requirements. This is a basic cleaning, preprocessing, and labeling process, and you may need to apply more complex techniques based on your data and task.

Also, remember to replace 'Your_Server_Name', 'Your_Database_Name', 'Your_User_ID', and 'Your_Password' with your actual SQL Server credentials.

Google Cloud Data Labeling Service can be used for more complex labeling tasks (e.g. if the labels can't be easily determined based on a simple rule). However, in this case, we can generate the labels directly in the DataFrame based on the threshold defined by the domain knowledge.

Step 3: Feature Engineering and Selection

Identify which characteristics or 'features' in your data are most relevant to the problem. This could be certain vibration patterns, frequencies, or other measures. Google Cloud's BigQuery ML can be used to create and select these features. Feature engineering and selection are crucial to improving the performance of your model.

Let's create some features from the frequency domain. It's common in vibration analysis to look at features like Peak (max amplitude), RMS (Root Mean Square), Frequency at max amplitude, Mean, and Standard Deviation.

Since we have three axes of measurement (let's call them X, Y, Z), we'll have to calculate these features for each axis. You can perform a Fast Fourier Transform (FFT) to move from the time domain to the frequency domain.

We can create additional features based on domain-specific knowledge about centrifugal pumps and the vibration patterns they produce under different conditions. For instance, we might be interested in the energy in certain frequency bands associated with specific types of mechanical faults.

Let's create a hypothetical scenario: based on the manufacturer's documentation, we expect that most mechanical faults will cause an increase in vibration at the pump's rotational frequency or its harmonics. Let's say the pump rotates at 30 Hz, so we're interested in the bands around 30 Hz, 60 Hz, 90 Hz, and so on.

We can implement our feature engineering function to compute each axis's energy in these frequency bands.

```python
def feature_engineering(df):
    # Perform FFT and calculate amplitude spectrum for each axis
    for axis in ['X', 'Y', 'Z']:
        # Calculate FFT
        fft_values = fft(df[axis])
        # Calculate amplitude spectrum
        amplitude_spectrum = np.abs(fft_values)
        df[f'fft_{axis}'] = amplitude_spectrum

        # Calculate features from the amplitude spectrum for each axis
        df[f'{axis}_peak'] = df[f'fft_{axis}'].apply(lambda x: np.max(x))  #
Peak
        df[f'{axis}_rms'] = df[f'fft_{axis}'].apply(lambda x:
np.sqrt(np.mean(np.square(x))))  # RMS
        df[f'{axis}_freq_at_peak'] = df[f'fft_{axis}'].apply(lambda x:
np.argmax(x))  # Frequency at max amplitude
        df[f'{axis}_mean'] = df[f'fft_{axis}'].apply(lambda x: np.mean(x))  #
Mean
        df[f'{axis}_std'] = df[f'fft_{axis}'].apply(lambda x: np.std(x))  #
Standard Deviation

        # Calculate energy in the bands of interest (rotational frequency and
its harmonics)
        for i in range(1, 5):  # up to the 4th harmonic
            freq_band = range(i * 30 - 5, i * 30 + 5)  # +/- 5 Hz around the
target frequency
            df[f'{axis}_energy_band_{i}'] = df[f'fft_{axis}'].apply(lambda x:
np.sum(x[freq_band]))

    return df

# Apply feature engineering to the DataFrame
df = feature_engineering(df)
```

Feature selection will be based on the correlation with the label:

```python
# Compute the correlation matrix
correlation_matrix = df.corr()

# Compute absolute value of correlation with the label
correlations_with_label = correlation_matrix['label'].apply(abs)

# Sort by descending correlation
sorted_correlations = correlations_with_label.sort_values(ascending=False)

# Set a correlation threshold for feature selection
correlation_threshold = 0.5

# Select features which have a correlation above the threshold
selected_features = sorted_correlations[sorted_correlations >
correlation_threshold].index

print("Selected features:")
print(selected_features)
```

Please note that the rotational frequency and bandwidths are hypothetical values for this example and should be adjusted according to the actual characteristics of the pump in question. Similarly, the correlation threshold for feature selection is a general guideline. It can be fine-tuned based on your specific requirements and resources.

Step 4: AI Model Selection and Development

Now, it's time to choose the machine learning model you'll use. You can use a pre-built AI model from Google AI Hub or develop a custom one using Google Cloud AI Platform. For predictive maintenance, you might choose a classification or a regression model based on the nature of your prediction task.

Since we are predicting a binary outcome (normal operation or mechanical failure), a classification model would be appropriate. Several types of models could be suitable, including decision trees, random forests, support vector machines, or neural networks, among others.

However, since we are dealing with time-series sensor data, you might consider using a model specifically designed for such data. A common choice for this is Long Short-Term Memory (LSTM), a Recurrent Neural Network (RNN) type that has proven effective for many time-series prediction tasks.

Here's how you might develop an LSTM model using TensorFlow, a popular machine-learning framework. This model will take the last 50 time steps of our multi-dimensional time-series data as input and output a binary prediction.

```python
from tensorflow.keras.models import Sequential
from tensorflow.keras.layers import LSTM, Dense, Dropout

# Split our data into training and validation sets
train_df = df.iloc[:int(len(df)*0.7)]
val_df = df.iloc[int(len(df)*0.7):]

# Scale the features (very important for neural networks)
from sklearn.preprocessing import StandardScaler

scaler = StandardScaler()
train_scaled = scaler.fit_transform(train_df.drop(columns='label'))
val_scaled = scaler.transform(val_df.drop(columns='label'))

# We need to reshape our data to be (samples, time steps, features)
X_train = train_scaled.reshape((train_scaled.shape[0], 1,
train_scaled.shape[1]))
X_val = val_scaled.reshape((val_scaled.shape[0], 1, val_scaled.shape[1]))

y_train = train_df['label']
y_val = val_df['label']

# Create a Sequential model
model = Sequential()
```

```python
model.add(LSTM(50, input_shape=(X_train.shape[1], X_train.shape[2]))) # 50 LSTM
units

model.add(Dropout(0.2)) # Dropout layer to prevent overfitting

model.add(Dense(1, activation='sigmoid')) # Binary output


# Compile the model
model.compile(loss='binary_crossentropy', optimizer='adam', metrics=['accuracy'])


# Train the model
history = model.fit(X_train, y_train, epochs=50, batch_size=64,
validation_data=(X_val, y_val), shuffle=False)
```

This will train the model for 50 epochs (iterations over the whole dataset) and validate it on the validation data. The model's performance should improve over time. However, suppose it begins to overfit (i.e., its performance on the training data continues improving while its performance on the validation data worsens). In that case, you may need to stop training early, adjust the model's complexity, or collect more data.

AI Model on Google Cloud

To use a pre-trained model, you first need to find a suitable model in the Google AI Hub. Here are the steps:
1. Visit the Google AI Hub.
2. Search for models related to predictive maintenance or time-series classification. Make sure the chosen model can handle your problem.
3. Once you select a model, follow the instructions to deploy it in your environment.

To build a custom model with Google Cloud AI Platform, the process involves the following steps:
1. First, you need to set up your environment, then upload your preprocessed and feature-engineered data to Google Cloud Storage.

```python
from google.cloud import storage

def upload_blob(bucket_name, source_file_name, destination_blob_name):
    """Uploads a file to the bucket."""
    storage_client = storage.Client()
    bucket = storage_client.get_bucket(bucket_name)
    blob = bucket.blob(destination_blob_name)
```

```python
    blob.upload_from_filename(source_file_name)

upload_blob('your-gcs-bucket', 'path/to/your/local/file',
'destination_blob_name')
```

2. Next, create a model using Google's AI Platform, where you define your machine
    learning model and specify the necessary settings. Here we'll use a simple decision
    tree classifier as an example:

```python
from google.cloud import aiplatform

aiplatform.init(project='your-gcp-project', location='us-central1')

# Create and deploy a model resource
model = aiplatform.Model.upload(
    display_name="predictive-maintenance-model",
    artifact_uri="gs://your-gcs-bucket/your-model/",
    serving_container_image_uri="us-docker.pkg.dev/cloud-
aiplatform/prediction/tf2-cpu.2-6:latest",
)

# Deploy the model
endpoint = model.deploy(machine_type="n1-standard-4")
```

**You might want to consider using Google's AutoML, which can automate much of this
process.**

Google AI Platform's Training service provides scalable resources for this. After the model is trained, validate its performance using a separate portion of your labeled data to evaluate the model's accuracy and generalizability. Google Cloud AI Platform's Evaluation service can be used for this purpose.

After selecting the features for your predictive maintenance system, the next step is to train and validate your model. You will first need to split your dataset into a training and validation set. Training the model involves feeding it your training data and letting it learn from it. Then the model's performance is validated using the validation set.

Let's say we've chosen to implement a Random Forest Classifier using Google's AI Platform. Below is a Python code example:

```python
from google.cloud import aiplatform
from sklearn.model_selection import train_test_split
from sklearn.ensemble import RandomForestClassifier
from sklearn.metrics import accuracy_score, confusion_matrix
import joblib
import os

# Assuming df is your DataFrame and 'failure' is the column that indicates a
failure
X = df.drop('failure', axis=1)  # The input features
y = df['failure']  # The target variable

# Split the data into training and validation sets (70% training, 30%
validation)
X_train, X_val, y_train, y_val = train_test_split(X, y, test_size=0.3,
random_state=42)

# Initialize a Random Forest classifier
clf = RandomForestClassifier(n_estimators=100)

# Train the model
clf.fit(X_train, y_train)

# Predict the validation set results
y_pred = clf.predict(X_val)

# Print accuracy score
print('Accuracy Score: ', accuracy_score(y_val, y_pred))

# Print confusion matrix
```

```python
print('Confusion Matrix: \n', confusion_matrix(y_val, y_pred))


# Save the model as a pickle file
joblib.dump(clf, 'model.pkl')


# Initialize Google Cloud Storage client
storage_client = storage.Client()


# Name of the bucket
bucket_name = "<Your_GCS_Bucket_Name>"


# Name of the blob
blob_name = "model.pkl"


# Create a blob in the bucket
blob = storage_client.get_bucket(bucket_name).blob(blob_name)


# Upload the model to GCS
blob.upload_from_filename('model.pkl')
```

This code splits the dataset into training and validation sets, trains a Random Forest classifier, makes predictions on the validation set, prints the accuracy score and confusion matrix, saves the model, and uploads it to a Google Cloud Storage bucket.

After training and validation, your model is ready to be deployed for making predictions. Be sure to evaluate the model's performance and adjust as needed. To fine-tune your predictive maintenance system, you might have to go back to data preprocessing, feature selection, or model selection steps.

Also, depending on the nature and complexity of your data, a different model might be more suitable. This trial-and-error process requires domain knowledge, an understanding of your data, and machine learning expertise.

Step 6: Implementation of the Predictive Model

Once your model is trained and validated satisfactorily, you can deploy it for making predictions. Implementing the predictive model involves analyzing new vibration data and predicting if a mechanical failure is imminent. This can be done using Google AI Platform's Predictions service.

Here's how you would go about this:

```python
from google.cloud import aiplatform

# Initialize the AI Platform
aiplatform.init(project='your-gcp-project', location='us-central1')

# Deploy your model (uploaded to a GCS bucket)
model = aiplatform.Model.upload(
    display_name="predictive-maintenance-model",
    artifact_uri="gs://your-gcs-bucket/your-model/",
    serving_container_image_uri="us-docker.pkg.dev/cloud-
aiplatform/prediction/tf2-cpu.2-6:latest",
)

# Deploy the model
endpoint = model.deploy(machine_type="n1-standard-4")
```

After deploying your model, you're ready to make predictions on new data:

```python
# Import the necessary modules
import googleapiclient.discovery
import json

# Define the name of your model
model_name = "projects/your-project/models/your_model"

# Initialize the AI Platform API
service = googleapiclient.discovery.build('ml', 'v1')

# Create a dictionary with the instances that you want to predict
instances = {
    'instances': [
        # Input your instance features
    ]
}
```

```python
# Create the request body
request_body = json.dumps(instances)


# Create the request
request = service.projects().predict(
    name=model_name,
    body=request_body
)


# Make the prediction
response = request.execute()

# Print the prediction
print(response)
```

This step involves turning the output of your AI model into useful actions. You can use Google Cloud Functions to translate the AI predictions into actionable tasks. This is a serverless execution environment where you can build and connect cloud services. You can create a function triggered when your model predicts a potential failure.

In this case, a possible action could be sending an email notification to the maintenance team or creating a maintenance task in a task management system. For this example, let's focus on sending an email notification.

First, you need to create a Google Cloud Function that sends an email when triggered. This can be done in JavaScript as Google Cloud Functions primarily support Node.js. Here's a sample function that sends an email using the SendGrid API:

```javascript
const sgMail = require('@sendgrid/mail');

exports.sendEmailNotification = (req, res) => {
  sgMail.setApiKey(process.env.SENDGRID_API_KEY);

  const msg = {
    to: 'maintenance_team@example.com',
    from: 'alert_system@example.com',
    subject: 'Maintenance Alert',
    text: 'The predictive maintenance system has detected a potential pump failure.',
  };

  sgMail.send(msg);

  res.status(200).send('Email sent');
};
```

This function would be deployed to Google Cloud and given a specific URL endpoint. You then trigger this function from your Python code when a potential failure is detected.

The following Python code makes a POST request to the Google Cloud Function when the model predicts a potential failure:

```python
import requests

predictions = loaded_model.predict(new_data)

# If the model predicts a failure, trigger Google Cloud Function
if any(predictions):
    response = requests.post('https://REGION-
PROJECT_ID.cloudfunctions.net/sendEmailNotification')

    # Check if the request was successful
    if response.status_code == 200:
        print('Maintenance alert sent successfully')
    else:
        print('Failed to send maintenance alert')
```

In the code above, replace 'REGION' and 'PROJECT_ID' with your Google Cloud region and project ID, respectively. This way, whenever your model predicts a potential failure, a POST request will be sent to the Google Cloud Function, triggering it to send an email alert to the maintenance team.

Remember to handle exceptions and errors when making the POST request. Always test the entire system thoroughly before moving to production to ensure everything works as expected.

Algorithmic Bias in AI and Mitigation Strategies

Algorithmic bias in artificial intelligence (AI) refers to the systematic and repeatable errors in a computer system that create unfair outcomes, such as privileging one arbitrary group of users over others. These biases can creep in through the data used to train the AI or the algorithms' design. For instance, if the training data predominantly represents one group, the AI may perform better for that group and may lead to biased outcomes for others.

Mitigating algorithmic bias involves multiple strategies. It begins with ensuring diversity in the data used to train AI systems. This includes a wide representation of different groups to avoid favoring any particular group. The next step involves developing robust, transparent algorithms. Transparency in AI algorithms allows their outcomes to be auditable and explainable, enabling the detection and correction of bias. Post-development bias-mitigation techniques can also be applied to check and correct for any observed bias in the AI's performance. Furthermore, it is crucial to have diverse teams developing and auditing these AI systems, as diversity in the human workforce can help to counteract unconscious biases that might seep into AI systems.

As AI continues to advance and become more integrated into our daily lives, we must address and correct for algorithmic bias to ensure fairness and equity in AI-driven decisions and outcomes.

Unveiling Algorithmic Bias

Algorithmic bias in physical asset management and maintenance can lead to inefficient operations, skewed predictive models, and an unsatisfactory level of service. AI and machine learning techniques are increasingly used in asset management for tasks ranging from predictive maintenance to condition monitoring and operational optimization. These algorithms learn from past data, which can lead to biased decisions if that data is biased.

For instance, if the data used to train a predictive maintenance model comes predominantly from newer assets or assets of a specific type, the model might need to perform better when predicting the maintenance needs of older or different types of assets. Similarly, if condition monitoring algorithms are trained on data from assets operating in certain environmental conditions, their performance may be compromised when applied to assets operating under different conditions.

Unveiling and addressing algorithmic bias in the context of physical asset management is crucial to ensure that the benefits of AI and machine learning are realized equitably across all assets. Strategies for mitigating algorithmic bias, such as diversifying training data and conducting regular bias audits, can help to promote fair and efficient asset management practices. As the field continues to embrace digital transformation, understanding and addressing algorithmic bias will be key to harnessing the full potential of AI in physical asset management and maintenance.

Algorithmic bias in asset management can arise from several causes, with significant consequences for the efficiency and fairness of asset management practices.

Causes of Algorithmic Bias

Data Bias: This is the most common cause of algorithmic bias. Suppose the data used to train the AI models needs to be more representative of the full range of assets or their operating conditions. In that case, the model will inherit these biases. For instance, if a predictive maintenance model is trained predominantly on data from one type of asset or newer assets, it may perform poorly when applied to other types of assets or older ones.

Selection Bias: This occurs when the data used to train the model is not randomly selected and does not represent the entire population of the data. For example, suppose data from multiple geographical regions is used to train the model. In that case, it may not accurately predict maintenance needs in other regions with different environmental conditions.

Confirmation Bias: This occurs when the model is influenced by pre-existing beliefs or expectations of the data scientists or engineers building the model. For example, suppose an engineer believes that a particular asset fails more frequently. In that case, they might unintentionally bias the model towards predicting more failures for that asset type.

Consequences of Algorithmic Bias

Inefficient Asset Management: Bias in AI models can lead to inaccurate predictions or inefficient recommendations, which can result in unnecessary maintenance activities, increased costs, and reduced asset lifetime.

Unfair Treatment of Assets: If the AI models are biased towards certain types of assets, other assets may not receive the necessary attention, resulting in premature failure or suboptimal operation.

Reduced Trust in AI Systems: If the stakeholders notice that the AI system's recommendations are consistently biased or inaccurate, they may lose trust in the system, which can hinder the adoption of AI in asset management.

Mitigating algorithmic bias involves

- ensuring diversity and representativeness in the training data,
- applying statistical methods to correct for known biases, and
- regularly auditing the performance of AI models across different asset types and conditions.

Organizations can ensure more efficient, fair, and trusted asset management practices by addressing algorithmic bias.

Algorithmic bias can seep into any system where algorithms and machine learning models are used, including asset maintenance and management in industries like wastewater utilities. Here are a few potential examples:

Predictive Maintenance Bias:

Consider a wastewater treatment facility using a machine learning algorithm for predictive maintenance. Suppose the algorithm was primarily trained on data from new, high-tech machinery. In that case, it might fail to predict maintenance needs for older equipment accurately. This could lead to an overemphasis on maintaining newer equipment. At the same time, older, potentially more failure-prone machinery might be neglected, resulting in unexpected breakdowns.

Geographic Bias:

An algorithm might be used to prioritize maintenance and upgrades in managing an extensive network of wastewater utilities. Suppose this algorithm is trained primarily on data from urban utilities. In that case, it might not perform as well when applied to rural utilities, potentially leading to an under-investment in these areas. This geographic bias could lead to disproportionate failures in rural utilities.

Vendor Bias:

Asset management often involves choosing between different equipment vendors. Suppose an algorithm used to assist this decision-making was trained primarily on data from one vendor's equipment. In that case, it might unfairly favor that vendor, even if another vendor's equipment would be a better fit. This vendor bias could lead to inefficient use of resources and potentially inferior asset performance.

Sensor Bias:

In a wastewater treatment plant, if the machine learning model for predicting equipment failure was trained on data from a specific type of sensor, it might perform poorly when analyzing data from a different type of sensor. This could lead to false positives or false negatives in failure prediction, leading to inefficiencies in maintenance scheduling.

Mitigating these biases is critical in ensuring fair and effective asset management. It involves diverse and representative data collection, transparent and interpretable algorithms, and regular evaluation of algorithmic performance across different scenarios. As of my knowledge cutoff in September 2021, discussions around algorithmic bias in AI are ongoing, and improvements in fairness and transparency in machine learning models are continually being developed.

Mitigating Algorithmic Bias in Asset Maintenance

In the digital transformation era, organizations across various sectors, including wastewater utilities, leverage Artificial Intelligence (AI) and Machine Learning (ML) to predict and optimize asset maintenance. However, these predictive models are not immune to bias. For example, the data used to train these models needs to be more representative. In that case, the resulting algorithmic decisions may also be biased, leading to inefficient maintenance practices and potentially costly downtime. This makes it crucial for organizations to mitigate algorithmic bias in their asset maintenance strategies actively. This process involves ensuring diverse and representative data collection, understanding and interpreting the mechanics of the algorithms and continuously monitoring and adjusting the system's performance. Organizations can ensure their asset maintenance operations are effective and equitable by prioritizing fairness and transparency in their algorithmic strategies.

Mitigating algorithmic bias is crucial to developing and implementing fair and reliable AI systems. Here are some strategies to mitigate algorithmic bias:

Diverse and Representative Data Collection:

One of the root causes of algorithmic bias is unrepresentative or biased training data. Collecting diverse and representative data can help ensure that the resulting AI models are fair and accurate. For instance, if you're developing an AI system for asset maintenance, ensure that the training data covers all types of assets from different manufacturers, ages, and operational conditions.

Data Preprocessing:

Preprocessing is another critical step in data preparation. It includes techniques like balancing the classes in the training data, handling missing or irrelevant data, and identifying potential bias in the data. Using these techniques can help prevent or minimize algorithmic bias.

Transparent and Interpretable Models:

Complex models may perform well, but their "black-box" nature can make identifying and correcting bias challenging. Instead, use interpretable models that allow you to understand how they're making predictions. This understanding can help identify potential sources of bias and make necessary adjustments.

Regular Audits:

Regularly auditing the performance of AI models is essential to identify and correct bias. These audits should be conducted on various dimensions – asset type, manufacturer, location, and other parameters, to ensure the model's predictions are accurate and unbiased across all categories.

Bias Mitigation Algorithms:

Various techniques and algorithms are designed to mitigate bias in AI models, such as adversarial debiasing, counterfactual fairness methods, or prejudice remover regularizers. Depending on the context and nature of the bias, these can be employed during the model training process.

Feedback Loop:

Establish a feedback loop to learn from the model's performance and errors continuously. This feedback can refine the model over time, improving its fairness and accuracy.

Stakeholder Participation:

Include diverse stakeholders in the process of algorithm development and auditing. Their perspectives can help identify potential bias and provide a more comprehensive understanding of fairness in the specific context of the application.

Legal and Ethical Guidelines:

Adhere to legal and ethical guidelines related to AI and algorithmic fairness. Regularly updating yourself with emerging research and regulations can help you stay compliant and ensure the ethical use of AI.

Organizations can mitigate algorithmic bias by integrating these strategies, leading to more equitable and reliable AI systems. This is beneficial from an ethical standpoint and can lead to improved performance and stakeholder trust in the system.

We now focus on practically implementing these concepts in a real-world scenario. In the previous chapter titled "From Theory to Practice: Implementing a Predictive Maintenance System for Pump Mechanical Failure Prediction," we discussed the step-by-step implementation of a predictive maintenance system. Looking back at this case, we can highlight where and how algorithmic bias might creep into our system, potentially affecting its performance and fairness. Using this predictive maintenance system as a backdrop, we will demonstrate how we can incorporate the strategies discussed earlier for mitigating algorithmic bias. Let's re-examine the journey from data collection to model deployment, this time focusing on identifying potential sources of bias and implementing strategies to counteract them.

Algorithmic bias in machine learning refers to systematic errors in the output of an algorithm that create unfair outcomes. This can be introduced in many stages of the machine learning pipeline, from biased data collection methods and the overrepresentation of certain classes in the data to the model's inherent assumptions about the data.
For instance, in our predictive maintenance system, if our training data is collected mostly during regular working hours (9am - 5pm), but the pump operates 24/7, the model may perform well during the day but poorly at night, because it's not well trained on data from those hours.

Similarly, suppose our labeled failure instances in the dataset mostly come from older pumps. In that case, the model might overestimate the likelihood of failure for older pumps and underestimate it for newer ones.

Mitigation strategies include:

- **Stratified Sampling**: To address potential bias in the data collection phase, ensure that the data you collect represents the different operating conditions of the pump. You can use stratified sampling to ensure that your data includes instances from different times of day, different days of the week, different pumps, etc.
- **Balanced Classes**: If your dataset has very few instances of failures compared to normal operation (which is likely the case), the model might become biased towards predicting normal operation. You can address this issue by oversampling the minority class (failures) or undersampling the majority class (normal operation) to balance the classes.

Here's an example of how you can balance your classes using the Synthetic Minority Over-sampling Technique (SMOTE) from the imbalanced-learn library:

```python
from imblearn.over_sampling import SMOTE

# separate features and labels
X = df.drop('failure', axis=1)  # assuming 'failure' is your target variable
y = df['failure']

# define SMOTE
sm = SMOTE(random_state=42)

# fit the sampling
X_res, y_res = sm.fit_resample(X, y)

# continue with your model training using the resampled data
```

- **Fairness Metrics**: Use fairness metrics to measure any potential bias after training your model. These can include metrics like disparate impact, equal opportunity difference, and more. Many Python libraries can help you compute these metrics, including Fairlearn and AIF360.

-

Mitigation strategies depend primarily on the nature of the bias. Understanding the potential sources of bias in your project is the first step toward addressing them.

Transparency and accountability are fundamental in combating bias, especially in artificial intelligence (AI) and algorithmic decision-making. Here's how they contribute to reducing bias:

Transparency:

Transparency refers to the openness and clear understanding of how the algorithm functions, makes decisions or predicts outcomes. This clarity is crucial for several reasons:

- Understanding Decision-Making: With transparency, developers, users, and stakeholders can understand how an AI system makes decisions. They can see which factors the algorithm considers significant and how they are used in decision-making.
- Detecting and Addressing Bias: When the workings of an algorithm are transparent, it becomes easier to detect bias. If an algorithm unfairly favors certain groups or outcomes, transparency can reveal this bias, allowing for corrective measures.
- Trust-Building: Transparency fosters trust. Users who understand how an AI system works and make decisions are more likely to trust its outputs.

Accountability:

Accountability in AI refers to the notion that organizations or individuals can be held responsible for the outcomes of the AI systems they deploy. It plays a crucial role in combating bias:

- Answerability: Holding parties accountable for their AI systems encourages them to take due care in their design, training, and testing processes, fostering an environment that prioritizes fairness and bias mitigation.
- Responsibility for Actions: When an AI system makes a biased or unjust decision, accountability means a system is in place to address this. This could involve rectifying the decision, adjusting the algorithm, or even providing compensation if harm has been caused.
- Incentivizing Fair Practices: Accountability can also serve as a deterrent against willfully ignoring or introducing bias into AI systems. Knowing they will be held responsible for biased outcomes can incentivize organizations to adopt rigorous bias mitigation strategies.

Therefore, transparency and accountability work hand-in-hand to reduce bias in AI and algorithmic systems. They help ensure that these systems are fair, just, and trustworthy, contributing to their effective and ethical use in various domains, including asset maintenance and management.

As AI continues to grow and evolve, several future directions are emerging in the field of mitigating algorithmic bias, particularly in the context of asset management:

Development of Bias-Aware Algorithms: Researchers are actively developing machine learning algorithms that are inherently bias-aware. These algorithms aim to reduce bias in their predictions by considering them during the learning process rather than correcting it afterward.

Explainable AI (XAI): The explainability and interpretability of AI models are key to understanding and mitigating algorithmic bias. Future developments in XAI can help to uncover hidden biases in AI models and provide more transparency in their decision-making process.

Automated Bias Detection Tools: Future advancements may see automated tools and software development that can detect and quantify bias in AI models. Such tools could greatly simplify the process of auditing AI systems for bias and help to ensure ongoing fairness and accuracy.

Regulatory Guidelines and Standards: As the impact of AI on society becomes significant, it's likely that more comprehensive regulatory guidelines and standards will be established to manage issues around algorithmic bias. These could include requirements for bias auditing, data collection standards, and transparency measures.

Ethics in AI: AI's ethical implications, including bias issues, are becoming a primary focus of discussion and research. This includes a greater emphasis on ethical training for AI practitioners and integrating ethical considerations into the AI development process.

Diversification of AI Teams: Ensuring that AI teams are diverse can also help to mitigate bias, as diverse teams bring different perspectives and are more likely to consider and identify potential biases. Future efforts may focus on promoting diversity and inclusion within the AI field.

By pursuing these and other strategies, we can work towards a future where algorithmic bias in asset management and other AI applications is effectively managed and mitigated.

The Role of Blockchain in Asset Management and Maintenance

Defining Blockchain Technology

Blockchain technology is a revolutionary digital concept that provides a method for validating transactions in a decentralized and secure manner. The technology was first conceptualized by an anonymous entity known as Satoshi Nakamoto in the Bitcoin white paper in 2008. Since then, it has been adapted for various applications beyond just cryptocurrency.

A blockchain is a chain of blocks, where each block represents a record of transactions. The unique aspect of a blockchain is that it is a distributed ledger, meaning the record of transactions is not held by a single central authority. However, it is distributed across a network of computers known as nodes. Each node in the network maintains a copy of the entire blockchain.

Key characteristics of blockchain technology include:
Decentralization: In a blockchain, transactions are recorded on a distributed ledger maintained by all network nodes. This means there is no single point of control or failure, which enhances the resilience and security of the system.

Immutability: Once a block is added to the blockchain, its information cannot be altered or deleted. This is because each block contains a unique code, known as a cryptographic hash, which depends on the block's information. Therefore, if the information is changed, the hash would also change, making it immediately apparent that the block has been tampered with.

Transparency: All transactions recorded on the blockchain are visible to all participants in the network. This transparency can foster trust among participants and reduce the risk of fraudulent activity.

Security: Transactions on a blockchain are secured using advanced cryptographic techniques. Before a block can be added to the chain, it must be validated by a majority of nodes in the network through a process known as consensus. This process ensures that only legitimate transactions are added to the blockchain.

Smart Contracts: These are self-executing contracts with the terms of the agreement being written into code. They automatically execute actions when pre-set conditions are met, automating various processes and transactions.

While blockchain technology holds great promise, it also comes with challenges such as scalability issues, energy consumption concerns (in the case of blockchains that use proof-of-work consensus mechanisms), and regulatory uncertainties. Despite these challenges, the technology's potential to disrupt traditional ways of conducting business and managing assets is widely recognized.

Blockchain technology, often associated with cryptocurrencies like Bitcoin, is increasingly recognized for its potential in asset management and maintenance. At its core, blockchain is a decentralized and immutable ledger of transactions distributed across a network of computers, also known as nodes. This innovative technology offers many advantages, such as enhanced security, transparency, and traceability, which are particularly beneficial in asset management.

The decentralized nature of blockchain means that all transactions and changes are stored in a way that is transparent to all participants in the network. This feature provides an indisputable record of asset history, including ownership, maintenance operations, and changes in condition. This transparency can significantly improve the accuracy of asset tracking and reduce the risk of fraud or error.
Moreover, the blockchain's immutable characteristic ensures that it can't be altered or deleted once a transaction or record has been added to the blockchain. This provides a reliable and secure record of each asset's lifecycle, crucial for effective asset maintenance. For instance, maintenance records stored on a blockchain could provide a verifiable history of repairs and inspections, making it easier to schedule future maintenance and predict potential issues.

In the context of asset management, blockchain could also streamline processes by enabling smart contracts – self-executing contracts with the terms of the agreement directly written into lines of code. These could automate various asset management tasks, such as triggering maintenance activities based on asset conditions or leasing and asset transfer transactions.

However, while the potential benefits are significant, the implementation of blockchain in asset management is still in its early stages, and many challenges need to be overcome, including technological complexity, interoperability issues, and regulatory considerations. Yet, as technology matures, it is expected to play an increasingly important role in asset management and maintenance.

Blockchain technology is poised to transform the asset management industry in several ways significantly:

Increased Transparency and Security: Blockchain technology provides an immutable, tamper-evident record of all transactions, creating unprecedented transparency. This transparency extends to all parts of the asset lifecycle, from acquisition to disposal, reducing the risk of fraud, misrepresentation, and dispute. Furthermore, using cryptography in blockchain ensures the security and integrity of data, making it a reliable source of truth for all involved parties.

Streamlined Processes: Blockchain can help automate and streamline several aspects of asset management, including compliance, auditing, and reporting. Smart contracts, or self-executing contracts with the terms directly written into code, can automate contract execution and enforcement, reducing the need for intermediaries and making transactions faster, cheaper, and more efficient.

Real-time Tracking: Blockchain technology can enable real-time tracking of assets, whether physical or digital. This can be especially useful in supply chain management, where the provenance and status of assets must be continuously monitored.

Tokenization of Assets: Blockchain allows for the tokenization of assets, turning real-world assets into digital tokens on the blockchain. This can make trading and managing assets easier, opening up new possibilities for fractional ownership and democratizing access to specific asset classes.

Improved Regulatory Compliance: The transparency and immutability of the blockchain can help asset managers to meet regulatory requirements more efficiently. The ability to quickly access and verify a complete, unalterable history of transactions can simplify audits and enhance regulatory reporting.

While these transformations are promising, it's important to note that the widespread adoption of blockchain in asset management also depends on regulatory acceptance, technological maturity, and overcoming challenges such as scalability and interoperability issues. Nevertheless, the potential impact of blockchain on the industry is significant and likely to unfold in the coming years.

Blockchain technology is poised to transform asset maintenance in several ways significantly:

Enhanced Traceability and Transparency: The immutable nature of blockchain makes it possible to maintain a reliable, tamper-proof history of an asset's maintenance events. This can increase the transparency and traceability of maintenance activities, making it easier to comply with regulatory requirements, resolve disputes, and manage warranties and service contracts.

Smart Contracts for Preventive Maintenance: With blockchain, preventive maintenance schedules can be encoded into smart contracts. These contracts automatically execute maintenance tasks when certain conditions are met, like reaching a specific date or hour of operation. This can streamline maintenance processes and ensure preventive maintenance is conducted promptly and accurately.

Real-time Monitoring and Decision Making: Blockchain can work with technologies like IoT and AI to facilitate real-time monitoring of asset conditions and performance. This data can be stored on the blockchain, providing a secure and reliable source of information for making maintenance decisions.

Enhanced Supply Chain Management: In situations where replacement parts are required, blockchain can improve supply chain transparency, helping to ensure that the right parts are available at the right time and place. This can reduce downtime and improve maintenance efficiency.

Credential Verification: Blockchain can also be used to verify the credentials of maintenance personnel, ensuring that only qualified individuals are allowed to perform specific maintenance tasks.

While the potential benefits of using blockchain in asset maintenance are significant, it's important to note that practical implementation of this technology also faces challenges, including technological complexity, interoperability issues, and the need for regulatory acceptance and standardization. Nonetheless, as the technology matures and these issues are addressed, blockchain will likely play an increasingly important role in asset maintenance.

Benefits:

Transparency and Traceability: Blockchain technology's immutable ledger offers complete transparency and traceability. This means each transaction and asset movement can be traced back to its origin, improving accountability and reducing the chances of fraud or asset misplacement. This enhanced traceability can improve the accuracy of maintenance records and help prevent disputes or errors.

Efficiency: Blockchain's decentralized nature can help reduce the need for intermediaries, resulting in faster and more efficient transactions.

Security: Blockchain networks use advanced cryptographic techniques to ensure that transactions are secure and data is tamper-proof, which can significantly enhance the security of asset management processes.

Smart Contracts: Blockchain technology enables smart contracts, which can automate various aspects of asset management, such as enforcing lease agreements, executing transactions when certain conditions are met, or triggering maintenance activities. Maintenance operations can be automated with smart contracts on the blockchain. These contracts can automatically trigger maintenance procedures when certain conditions are met, reducing the possibility of human error.

Secure Record Keeping: Blockchain's cryptographic security measures ensure that maintenance records cannot be tampered with. This provides a reliable and secure record of all maintenance activities for each asset.

Real-time Information Sharing: Blockchain can facilitate the real-time sharing of maintenance data across a network. This can help coordinate maintenance activities more efficiently and ensure that all relevant parties can access the most up-to-date information.

Integration with Other Technologies: Blockchain can be integrated with other technologies, such as the Internet of Things (IoT) and artificial intelligence (AI), to provide real-time asset tracking, predictive maintenance, and data-driven decision-making. Blockchain can be integrated with Internet of Things (IoT) devices for real-time monitoring of asset conditions. This can enable predictive maintenance, where potential problems are identified and addressed before they result in asset failure.

Complexity and Technical Knowledge: Implementing blockchain technology requires specific technical expertise, which might be challenging for organizations without a strong IT department.

Regulation and Compliance: Blockchain is a relatively new technology, and the regulatory environment around its use in asset management is still evolving. This can create uncertainty and risk for businesses.

Scalability: While blockchain can handle many transactions, as the number of users and transactions increases, so does the computational power required. As the number of assets and associated maintenance activities increases, the computational resources required for maintaining the blockchain network can become significant. This scalability issue can limit the applicability of blockchain in larger or more complex asset maintenance operations.

Adoption and Change Management: As with any new technology, blockchain requires overcoming resistance to change. This includes educating staff about blockchain and re-engineering business processes to exploit the technology fully.

Privacy Concerns: When sensitive data is involved, blockchain can enhance transparency and raise privacy concerns. Balancing transparency with privacy when implementing blockchain in asset management or maintenance operations is essential.

Lack of Standardization and Interoperability: There are many different blockchain platforms, and not all of them are compatible with each other. This lack of standardization and interoperability can limit the usefulness of blockchain for asset maintenance.

Introduction to Smart Contracts:

Smart contracts are self-executing contracts with the terms of the agreement directly written into code. They exist across a distributed, decentralized blockchain network and automatically execute transactions once predetermined conditions are met.
Smart contracts, developed with blockchain technology, encode business rules in a programmable language. This allows for automation, precision, and impartiality not typically possible with traditional contracts. Once the conditions stipulated in the contract are met, the actions specified in the agreement are automatically triggered and carried out.

For example, in the context of asset management and maintenance, a smart contract could be programmed to automatically order a replacement part when a piece of machinery reaches a certain number of operating hours or shows signs of wear and tear. This could streamline maintenance, minimize downtime, and improve operational efficiency.

Despite their potential advantages, smart contracts also come with their own set of challenges. Understanding the technical language of smart contracts requires a certain level of proficiency in coding, which may not be universally accessible. Additionally, as they are immutable once deployed, any bugs or errors in the code can lead to unintended consequences. Security is also a concern, as malicious actors can exploit vulnerabilities in smart contracts. Nonetheless, with careful design and implementation, smart contracts hold tremendous potential for transforming business operations in numerous sectors, including asset management and maintenance.

Smart contracts, which are programmable contracts that automatically execute when predefined conditions are met, have the potential to transform various aspects of asset management. These are some of the ways they can be beneficial:

Automation of processes: Many asset management processes can be automated using smart contracts. For example, when it comes to maintenance schedules, a smart contract could automatically trigger a maintenance request when certain conditions related to an asset's usage or performance are met. This could streamline the maintenance process, reducing downtime and enhancing efficiency.

Improved transparency: Because smart contracts are stored on a blockchain, they provide a transparent and immutable record of transactions. This can enhance trust among parties, reduce disputes, and provide a clear audit trail.

Reduced intermediaries and costs: Smart contracts can execute transactions automatically without intermediaries, leading to cost savings. This could be particularly beneficial in asset management, where transactions such as procurement, leasing, or disposal of assets often involve multiple parties and complex contractual arrangements.

Enhanced security: Blockchain technology's decentralized nature enhances the security of smart contracts. This makes them less prone to hacking, fraud, and other forms of manipulation, providing an additional layer of security in asset management.

Regulatory compliance: Smart contracts can be programmed to adhere to regulatory requirements, ensuring that all transactions comply with the relevant regulations. This can simplify the regulatory compliance process and reduce the risk of non-compliance penalties.

Despite the potential benefits, there are also challenges in implementing smart contracts in asset management. These include the technical complexity of programming smart contracts, potential privacy and data protection issues, and the current need for standardized legal frameworks and regulations governing smart contracts.

This chapter will delve into how blockchain can enhance the management of asset lifecycles by using the Hyperledger Fabric platform, a popular and highly versatile open-source blockchain framework. The chapter will illustrate a step-by-step implementation guide using a wastewater pump as the sample asset.

Given the depth and technical nature of the subject, we decided to focus this guide on the pump's installation data as a starting point, establishing a foundation for managing its entire lifecycle eventually. The selected blockchain technology is Hyperledger Fabric due to its versatility, robust security features, and a high degree of control over variables like confidentiality, resilience, and flexibility.

Hyperledger Fabric(https://www.hyperledger.org/), part of the Linux Foundation's Hyperledger project, is a distributed ledger platform that is highly modular and configurable, making it an ideal choice for a wide variety of applications. Adopting Hyperledger Fabric for asset lifecycle management provides numerous benefits as we venture into the public utility services sector.

- **Permissioned Network**: Unlike public blockchains, Hyperledger Fabric is a permissioned blockchain, meaning that participants are known and identifiable. This is crucial for a public utility service like a wastewater treatment facility where security and regulatory compliance are paramount.
- **Scalability and Performance**: Hyperledger Fabric's architecture is designed for scalability and performance. Its unique execute-order-validate architecture, where the transaction execution is separated from consensus, allows for high transaction throughput and low latency, enabling real-time asset lifecycle management.
- **Privacy and Confidentiality**: Hyperledger Fabric offers enhanced privacy and confidentiality features, allowing different organizations to maintain separate ledgers and run smart contracts without sharing sensitive information. This is especially important for public service utilities, which often have strict data protection requirements.
- **Modularity**: Hyperledger Fabric's modular architecture allows for plug-and-play components, enabling customization according to specific requirements of the public service utility sector.
- **Interoperability**: Hyperledger Fabric's interoperability with other systems (like the CMMS system in this case) allows seamless integration of blockchain technology into existing infrastructures, minimizing disruptions and maximizing efficiency.

Given these advantages, Hyperledger Fabric is ideal for implementing blockchain technology in the public service utility sector, particularly for asset lifecycle management. Let's now explore the step-by-step implementation of this powerful solution.

Transition to Implementation Guide

Before we delve into the implementation details, it's important to understand the overall structure of the solution we are about to build. We will start with a system design and architecture, outlining the key components of a Hyperledger Fabric blockchain network. Following this, we will define our asset and the transaction type, which in this case, is pump installation information.

The core logic of our solution will reside in the chaincode, also known as the smart contract. Despite the notion that our system doesn't necessitate a smart contract, it's crucial to understand that chaincode is essential for defining the business rules of the blockchain. We'll elucidate the chaincode functions needed for our system.

Next, we will look at how to develop an API for seamless interaction between the CMMS system and the blockchain. Subsequently, we will provide detailed steps for integrating the solution with the existing CMMS system using the newly created API. The guide will conclude with a section on testing and validation, ensuring that the solution operates correctly and as intended.

With our chosen example, the pump, let's dive into the fascinating world of blockchain and explore how it can transform how we manage and maintain asset lifecycles.

Step-by-Step Guide to Implementing Hyperledger Fabric Blockchain for Pump Lifecycle Management

We're operating under several assumptions in this implementation guide to streamline the process. We're focusing on a single organization named "XYZ Wastewater Treatment Services." As such, all the procedures, nomenclatures, and defaults will be tailored specifically to this organization and its operations. We also assume that the system's end users will be the employees of XYZ Wastewater Treatment Services, who will interact with the blockchain to manage the lifecycle of the assets.

Speaking of assets, we've pre-selected a pump as the central asset for this guide. All the data defaults, names, and values used throughout the guide will be chosen with respect to the pump. This approach helps create a specific, relatable, and practical guide, allowing us to provide a detailed, step-by-step journey for implementing blockchain technology for pump lifecycle management within the context of XYZ Wastewater Treatment Services.

Step 1: System Architecture and Design

1.1 Define the Network:

Before getting into the implementation, let's visualize the architecture of our network. We will have one organization: XYZOrg. This organization will maintain a single peer node to store the ledger and the chaincode.

This is a simple setup of a Hyperledger Fabric network.

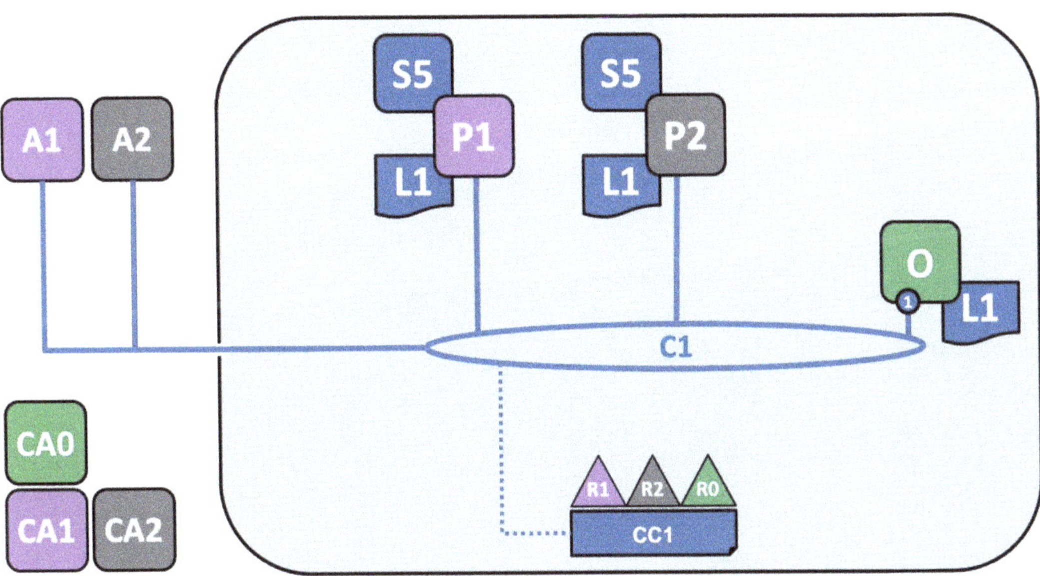

Now, let's move towards setting up the network.

1.2 Setting up the Network

Here are the steps to setup a Hyperledger Fabric network:

Install Prerequisites

- Linux
- Docker and Docker Compose
- Node.js and npm
- Go Programming Language
- Python

You can find detailed installation guides for these prerequisites in the official Hyperledger Fabric documentation(https://hyperledger-fabric.readthedocs.io/en/release-2.2/prereqs.html).

1.3 Install Hyperledger Fabric Samples, Binaries, and Docker Images

Navigate to your workspace and run the following command:

```bash
curl -sSL https://bit.ly/2ysbOFE | bash -s
```

This command downloads and executes a bash script that will download and extract all of the platform-specific binaries needed to setup your network and place them into the bin sub-directory of the current working directory.

1.4 Set Up the Network Artifacts:

The next step involves creating the necessary configuration files that define our network, i.e., the organization, peer node, ordering service, etc.

This is usually done through a **configtx.yaml** file. Here's a simple sample for the XYZOrg:

```yaml
Organizations:

    - &XYZOrg
        Name: XYZOrgMSP
        ID: XYZOrgMSP
        MSPDir: crypto-config/peerOrganizations/xyzorg.example.com/msp
        AnchorPeers:
            - Host: peer0.xyzorg.example.com
              Port: 7051

Application: &ApplicationDefaults
    Organizations:

Orderer: &OrdererDefaults
    OrdererType: solo
    Addresses:
        - orderer.example.com:7050
```

```yaml
Profiles:

    XYZOrgGenesis:

        <<: *ChannelDefaults

        Orderer:

            <<: *OrdererDefaults

            Organizations:

                - *XYZOrg

        Consortiums:

            SampleConsortium:

                Organizations:

                    - *XYZOrg


    XYZOrgChannel:

        Consortium: SampleConsortium

        <<: *ChannelDefaults

        Application:

            <<: *ApplicationDefaults

            Organizations:

                - *XYZOrg
```

This file defines a single organization, XYZOrg, with one peer node.

1.4 Generate the Network Artifacts:

Using the configtx.yaml file, we now generate the necessary artifacts:

```
./bin/configtxgen -profile XYZOrgChannel -outputCreateChannelTx ./channel-
artifacts/channel.tx -channelID xyzchannel

./bin/configtxgen -profile XYZOrgChannel -outputAnchorPeersUpdate ./channel-
artifacts/XYZOrgMSPanchors.tx -channelID xyzchannel -asOrg XYZOrgMSP
```

Once you've created the necessary network artifacts, the final step is to use Docker to bring your network online. This involves creating a Docker Compose file, typically named **docker-compose.yaml**, that will describe and configure all of your network's components.

You can bring up the network by navigating to the directory containing the docker-compose.yaml file and running the following command:

```bash
docker-compose up -d
```

This command starts all the services described in your Docker Compose file in the background. The **-d** flag stands for "detached mode".

To confirm that your network is up and running, you can list all running Docker containers with:

```bash
docker ps
```

This should show you the running containers for your peer nodes and ordering service. This command should output the list of the currently running Docker containers. If your network is running, you should see your peer and orderer in the list.

Remember, this is a simplified network setup intended for a development setting. In a production environment, additional configurations and security measures would be needed.

Note: Always check the latest Hyperledger Fabric documentation as the setup process might have changed slightly.

Next, we need to join our peer node to the channel.

```bash
./network.sh deployCC -ccn basic -ccp ../chaincode/basic/ -ccl go
```

The flags in this command do the following:
- deployCC - deploys the chaincode.
- -ccn basic - sets the name of the chaincode to basic.
- -ccp ../chaincode/basic/ - sets the path to the chaincode.
- -ccl go - sets the language of the chaincode to Go.

We now have an organization called XYZOrg with one peer node. The peer is part of a channel named mychannel, and a chaincode named basic has been deployed to the peer.

1.7 Identity Management

Identity management in a Hyperledger Fabric network is handled through the Membership Service Providers (MSPs), which authenticate, validate, and manage identities on the network. We'll set up our own Certificate Authority (CA) for XYZOrg, which will issue and validate certificates to control network access for XYZ Wastewater Treatment Services employees.

Setup Certificate Authority

The network.sh script we used in the previous step also starts a Certificate Authority (CA) for you and created an admin identity for each organization.

1. To enroll an admin user and generate crypto material for them, navigate to the test-network directory and execute:

```bash
./network.sh registerUser admin
```

2. To create an employee user, execute the following command:

```bash
./network.sh registerUser john
```

This will generate a user named 'john'. You can replace 'john' with any name you prefer.

After you've registered users, their credentials are stored in the wallet directory.

Implementing Identity Management in Code

For the code to interact with the network, it needs to use these identities. Here's an example using Node.js.

First, install the necessary package:

```bash
npm install fabric-network
```

Then, you can use the following script to connect to the network as user 'john', gets a reference to the chaincode 'basic,' and invokes the 'getAllAssets' function of the chaincode. It then logs the results of the transaction.

```javascript
const { Gateway, Wallets } = require('fabric-network');
const path = require('path');

async function main() {
    try {
        // load the network configuration
        const ccpPath = path.resolve(__dirname, '..', '..', 'test-network',
'organizations', 'peerOrganizations', 'org1.example.com', 'connection-
org1.json');

        // Create a new file system based wallet for managing identities.
        const walletPath = path.join(process.cwd(), 'wallet');
        const wallet = await Wallets.newFileSystemWallet(walletPath);

        // Check to see if we've already enrolled the user.
        const identity = await wallet.get('john');
        if (!identity) {
            console.log('An identity for the user "john" does not exist in the
wallet');
            console.log('Run the registerUser.js application before retrying');
            return;
        }

        // Create a new gateway for connecting to our peer node.
        const gateway = new Gateway();
        await gateway.connect(ccpPath, { wallet, identity: 'john', discovery: {
enabled: true, asLocalhost: true } });

        // Get the network (channel) our contract is deployed to.
        const network = await gateway.getNetwork('mychannel');
```

```javascript
        // Get the contract from the network.

        const contract = network.getContract('basic');


        // Evaluate the specified transaction.

        const result = await contract.evaluateTransaction('getAllAssets');

        console.log(`Transaction has been evaluated, result is:
${result.toString()}`);


    } catch (error) {

        console.error(`Failed to evaluate transaction: ${error}`);

        process.exit(1);

    }

}

main();
```

Remember to replace 'john' and 'basic' with the names you've chosen for the user and chaincode respectively. Also, you should have a function 'getAllAssets' in your chaincode. This is a basic example, but you can create identities for as many users as you need in a similar fashion.

Step 2: Digital Representation of a Pump

In this step, we create a digital twin for the pump that will be used for tracking on the blockchain. The digital twin is effectively a detailed digital model of the physical asset - in this case, the pump - which carries essential attributes such as Pump ID, Model, Manufacturer, Installation Date, and Location.

We will represent a digital twin of the pump on the Hyperledger Fabric blockchain using chaincode, written in Node.js. The chaincode encapsulates the business logic of the blockchain application and maintains the state of the pump as an asset on the ledger.
Here's a simple Node.js example showing how to represent the pump:

```javascript
'use strict';

const { Contract } = require('fabric-contract-api');

class Pump extends Contract {

    async pumpExists(ctx, pumpId) {
        const buffer = await ctx.stub.getState(pumpId);
        return (!!buffer && buffer.length > 0);
    }

    async createPump(ctx, pumpId, model, manufacturer, installationDate,
location) {
        const exists = await this.pumpExists(ctx, pumpId);
        if (exists) {
            throw new Error(`The pump ${pumpId} already exists`);
        }
        const asset = { model, manufacturer, installationDate, location };
        const buffer = Buffer.from(JSON.stringify(asset));
        await ctx.stub.putState(pumpId, buffer);
    }

    async readPump(ctx, pumpId) {
        const exists = await this.pumpExists(ctx, pumpId);
        if (!exists) {
            throw new Error(`The pump ${pumpId} does not exist`);
        }
        const buffer = await ctx.stub.getState(pumpId);
        const asset = JSON.parse(buffer.toString());
        return asset;
    }
}
```

```
module.exports = Pump;
```

This Node.js chaincode represents a pump with your specified properties. It has a **pumpExists** function for checking if a pump installation exists, a **createPump** function for creating new pump installations, and a **readPump** function to read a pump's details from the ledger.

Step 3: Define the Blockchain Model

We'll use Hyperledger Fabric's chaincode to define the blockchain model. The chaincode will model the pump asset and its installation transactions. This model will be used for tracking the pump on the blockchain.

The model can be designed as follows using Node.js:

```javascript
'use strict';

const { Contract } = require('fabric-contract-api');

class Pump extends Contract {

    async pumpExists(ctx, pumpId) {
        const buffer = await ctx.stub.getState(pumpId);
        return (!!buffer && buffer.length > 0);
    }


    async createPump(ctx, pumpId, model, manufacturer, installationDate,
location) {
        const exists = await this.pumpExists(ctx, pumpId);
        if (exists) {
            throw new Error(`The pump ${pumpId} already exists`);
        }
        const asset = { model, manufacturer, installationDate, location };
        const buffer = Buffer.from(JSON.stringify(asset));
        await ctx.stub.putState(pumpId, buffer);
    }


    async readPump(ctx, pumpId) {
        const exists = await this.pumpExists(ctx, pumpId);
        if (!exists) {
            throw new Error(`The pump ${pumpId} does not exist`);
        }
        const buffer = await ctx.stub.getState(pumpId);
        const asset = JSON.parse(buffer.toString());
        return asset;
    }
}

module.exports = Pump;
```

This Node.js chaincode represents a pump with the specified properties (Pump ID, model, manufacturer, installation date, and location). It has a **pumpExists** function to check if a pump installation exists, a **createPump** function to create new pump installations, and a **readPump** function to read a pump's details from the ledger. These functions will be invoked when transactions are processed on the network.

While defining the blockchain model, it's also important to remember the immutability property of the blockchain. Once the pump installation information has been recorded, it can't be modified or deleted. Any updates would have to be recorded as new transactions.

Step 4: Chaincode (Smart Contract) Development

Chaincode, also known as Smart Contract, is essentially the business logic of a blockchain network. It defines how participants can interact with the ledger and contains the rules agreed by the members of the network.

For your pilot project, we will write a simple chaincode that allows us to create a new pump installation and retrieve the details of an existing one. For the sake of this example, I'll be using Node.js.

```javascript
'use strict';

const { Contract } = require('fabric-contract-api');

class PumpContract extends Contract {

    async initLedger(ctx) {
        console.info('============== START : Initialize Ledger ============');
        const pumps = [
            {
                model: 'Model 1',
                manufacturer: 'Manufacturer 1',
                installationDate: '01/01/2022',
                location: 'Location 1',
            },
            // Add more pumps as needed
        ];

        for (let i = 0; i < pumps.length; i++) {
            pumps[i].docType = 'pump';
            await ctx.stub.putState('PUMP' + i,
Buffer.from(JSON.stringify(pumps[i])));
            console.info('Added <--> ', pumps[i]);
        }
        console.info('============== END : Initialize Ledger ============');
    }

    async createPump(ctx, pumpId, model, manufacturer, installationDate,
location) {
        console.info('============== START : Create Pump ============');

        const pump = {
            docType: 'pump',
            model,
            manufacturer,
            installationDate,
```

```javascript
        location,
    };

    await ctx.stub.putState(pumpId, Buffer.from(JSON.stringify(pump)));
    console.info('============= END : Create Pump ===========');
}


async queryPump(ctx, pumpId) {
    const pumpAsBytes = await ctx.stub.getState(pumpId);
    if (!pumpAsBytes || pumpAsBytes.length === 0) {
        throw new Error(`${pumpId} does not exist`);
    }
    console.info(pumpAsBytes.toString());
    return pumpAsBytes.toString();

    }
}

module.exports = PumpContract;
```

The **initLedger** function initializes the ledger with some pumps for testing purposes. The **createPump** function creates a new pump with given details. The **queryPump** function fetches the details of a pump by its ID.

Make sure to replace the placeholders in the pump object (in the **createPump** function) with actual data before running this code.

To install and instantiate the chaincode, we would typically use the **peer** CLI tool provided by Hyperledger Fabric, along with some commands in a terminal. However, For this guide, I'll be using the Hyperledger Fabric SDK for Node.js to demonstrate this.

Here is an example of how you can install and instantiate the chaincode using Node.js:

```javascript
const { FileSystemWallet, Gateway, X509WalletMixin } = require('fabric-
network');
const FabricCAServices = require('fabric-ca-client');
const path = require('path');
const fs = require('fs');

// Load connection profile
const ccpPath = path.resolve(__dirname, '..', '..', 'network', 'connection-
profile.json');
const ccpJSON = fs.readFileSync(ccpPath, 'utf8');
const ccp = JSON.parse(ccpJSON);

// Setup wallet
const walletPath = path.join(process.cwd(), 'wallet');
const wallet = new FileSystemWallet(walletPath);

// Admin identity
const adminIdentity = X509WalletMixin.createIdentity('XYZOrgMSP',
'Admin@XYZOrg-cert.pem', 'Admin@XYZOrg-priv.pem');
await wallet.import('Admin@XYZOrg', adminIdentity);

// Connect to gateway
const gateway = new Gateway();
await gateway.connect(ccp, { wallet, identity: 'Admin@XYZOrg', discovery: {
enabled: false } });

// Get admin
const admin = gateway.getCurrentIdentity();

// Connect to XYZOrg peer
const client = gateway.getClient();
const peer = client.getPeer('peer0.XYZOrg.example.com');

// Install chaincode
const chaincodePath = path.resolve(__dirname, '..', '..', 'chaincode',
'pumpContract.js');
const installRequest = {
    targets: [peer],
```

```javascript
    chaincodePath,
    chaincodeId: 'pumpContract',
    chaincodeVersion: '1.0',
    chaincodeType: 'node',
};
await client.installChaincode(installRequest);

// Instantiate chaincode
const channel = client.getChannel('mychannel');
const instantiateRequest = {
    targets: [peer],
    chaincodeId: 'pumpContract',
    chaincodeVersion: '1.0',
    args: ['initLedger'],
    chaincodeType: 'node',
    txId: client.newTransactionID(true),
};
const proposalResponses = await
channel.sendInstantiateProposal(instantiateRequest);
const proposal = proposalResponses[1];
const response = proposalResponses[0];

if (response.response.status === 200) {
    const request = {
        proposalResponses,
        proposal,
    };
    const txId = client.newTransactionID(true);
    const eventPromises = [];
    eventPromises.push(channel.sendTransaction(request, txId));
    await Promise.all(eventPromises);
}
```

Please note that this is a simplification. The actual code might vary depending on your specific use case. You'll also need to replace placeholders (like **Admin@XYZOrg-cert.pem** and **Admin@XYZOrg-priv.pem**) with actual paths to your admin certificate and private key files. Ensure your connection profile, wallet, and chaincode files are located correctly relative to your script.

Here's how you can create a simple RESTful API using Node.js and Express that interacts with the blockchain network via the Fabric SDK. We'll expose two endpoints - one for creating a new pump installation and one for fetching a pump installation.

```
const express = require('express');
const { FileSystemWallet, Gateway } = require('fabric-network');
const path = require('path');
const fs = require('fs');

const app = express();
app.use(express.json());

const ccpPath = path.resolve(__dirname, '..', '..', 'network', 'connection-
profile.json');
const ccpJSON = fs.readFileSync(ccpPath, 'utf8');
const ccp = JSON.parse(ccpJSON);

const walletPath = path.join(process.cwd(), 'wallet');
const wallet = new FileSystemWallet(walletPath);

app.post('/api/installation', async (req, res) => {
    const gateway = new Gateway();
    await gateway.connect(ccp, { wallet, identity: 'Admin@XYZOrg', discovery: {
enabled: false } });

    const network = await gateway.getNetwork('mychannel');
    const contract = network.getContract('pumpContract');

    const result = await contract.submitTransaction('createPumpInstallation',
JSON.stringify(req.body));
    res.status(200).json({ result: result.toString() });

    await gateway.disconnect();
});

app.get('/api/installation/:id', async (req, res) => {
    const gateway = new Gateway();
    await gateway.connect(ccp, { wallet, identity: 'Admin@XYZOrg', discovery: {
enabled: false } });

    const network = await gateway.getNetwork('mychannel');
    const contract = network.getContract('pumpContract');
```

```javascript
    const result = await contract.evaluateTransaction('getPumpInstallation',
req.params.id);
    res.status(200).json({ result: JSON.parse(result.toString()) });

    await gateway.disconnect();
});

app.listen(3000, () => {
    console.log('Server running on port 3000');
});
```

In the above code:

- We define an Express app and use the built-in **express.json()** middleware to parse incoming JSON payloads.

- We then create two endpoints:

    - A **POST** endpoint at **/api/installation** that accepts pump installation data, invokes the **createPumpInstallation** transaction on our smart contract, and returns the transaction result.

    - A **GET** endpoint at **/api/installation/:id** that accepts a pump ID as a URL parameter, invokes the **getPumpInstallation** transaction on our smart contract, and returns the queried pump installation data.

- We connect to the Hyperledger Fabric network using the **Gateway** class provided by the Fabric SDK. To do so, we need to point it to a connection profile (in JSON format) and provide a wallet containing the identities authorized to interact with the network.

- We connect to a specific network channel and retrieve our smart contract using the **getNetwork** and **getContract** methods, respectively. After that, we either **submitTransaction** (which modifies the ledger and therefore needs to be endorsed and committed) or **evaluateTransaction** (which queries data and therefore does not modify the ledger).

- Finally, we start our server by listening on port 3000.

Step 7: Integration with the CMMS System

Define the process for the CMMS system to interact with the blockchain API. A scheduled job might be used to record new pump installations.

Step 8: Testing and Validation

Finally, we need to conduct thorough testing to ensure the correct functioning of the blockchain system, validating that the CMMS system interacts correctly with the blockchain via the API, and verifying the correct recording and retrieval of pump installation data.

As we move further into the digital age, the potential applications and impact of artificial intelligence (AI) and blockchain technology in asset management are poised to be transformative. These technologies are expected to evolve and be crucial in driving the sector's efficiency, transparency, security, and automation.

The future of AI in asset management is expected to be marked by more sophisticated predictive and cognitive maintenance systems. These systems will likely predict asset failures with even greater accuracy and far in advance, enabling organizations to implement proactive maintenance strategies and thus reduce downtime and costs. With advancements in machine learning and data analytics, AI will also be more adept at identifying patterns and trends, providing valuable insights for decision-making.

On the blockchain front, we anticipate seeing more widespread adoption of this technology in asset management, particularly smart contracts. These programmable contracts have the potential to automate numerous transactional and contractual processes, significantly increasing efficiency. Blockchain's immutable and transparent nature could also boost trust and security in asset transactions and records.

Moreover, the convergence of AI and blockchain could usher in even more advanced solutions. For example, AI could be used to analyze data stored on a blockchain, providing enhanced insights and decision-making capabilities. Alternatively, blockchain could be used to secure AI models and algorithms, ensuring their integrity.

However, these future developments will not be without their challenges. Organizations will need to overcome issues related to data privacy, security, algorithmic bias, and the technical complexities of implementing and managing these advanced technologies. The regulatory landscape must also evolve to accommodate these new technologies and their applications. Despite these hurdles, AI and blockchain's potential benefits to asset management are immense. They will likely drive continued innovation and adoption in the sector.

As AI and blockchain continue to evolve, several emerging trends are worth noting:

Integration of AI and Blockchain: As these two technologies mature, we see more instances of being used together. AI can help analyze and interpret the vast amounts of data stored on a blockchain, making it more valuable and accessible. At the same time, blockchain can provide a secure, unalterable platform for AI algorithms and models, ensuring their reliability and trustworthiness.

Development of Decentralized AI: There is an increasing interest in developing decentralized AI models, leveraging blockchain's decentralized nature. This approach could help democratize AI, allowing multiple parties to contribute to and benefit from AI models while maintaining transparency and trust.

AI-driven Smart Contracts: Smart contracts on blockchain platforms are becoming more sophisticated with the integration of AI. This can enable more complex, automated transactions that can respond dynamically to changes in the data or environment.

Blockchain for Data Governance in AI: With growing concerns about data privacy and security in AI, blockchain is being explored as a solution for better data governance. It can provide a secure, transparent, and decentralized system for managing and tracking data used in AI, ensuring compliance with regulations and user consent.

AI for Blockchain Scalability and Efficiency: Blockchain faces scalability and efficiency issues, especially in large-scale applications. AI and machine learning algorithms are being used to optimize blockchain operations, improve scalability, and reduce energy consumption.

Autonomous Decentralized Organizations: With the combination of AI and blockchain, there is potential for creating Autonomous Decentralized Organizations (DAOs). These are self-governing, largely automated organizations that operate based on pre-set rules encoded as smart contracts. AI can aid in decision-making processes within these organizations.

Quantum Computing and Blockchain: As quantum computing progresses, it is expected to impact AI and blockchain. Quantum computing could significantly speed up AI processes and secure blockchain against attacks. However, it also presents new challenges and risks.

While these trends are promising, they also come with challenges like technical complexity, regulatory concerns, and ethical considerations. Nonetheless, the potential benefits of integrating AI and blockchain could drive significant innovation and transformation in various sectors, including asset management.

The convergence of AI and blockchain creates transformative asset maintenance and management possibilities. This fusion is expected to bring about unprecedented efficiency, security, transparency, and autonomy in the management of assets, particularly in the context of physical infrastructure. Here's how:

Enhanced Decision-making and Predictive Capabilities: In asset management, AI can significantly enhance decision-making processes by analyzing vast amounts of data, predicting future scenarios, and leveraging its ability. For instance, AI-powered predictive maintenance can help prevent equipment failures, reducing downtime and maintenance costs. When coupled with blockchain's ability to ensure the authenticity and immutability of data, the decisions made based on AI's insights become even more reliable and trustworthy.

Increased Transparency and Traceability: Blockchain's distributed ledger technology provides an unalterable, time-stamped record of all transactions and activities, increasing transparency and traceability. This can be particularly beneficial in asset management, where tracking the lifecycle of assets, from acquisition to disposal, is crucial. AI can further enhance this by intelligently analyzing and presenting blockchain-stored data, making it easier for stakeholders to understand and use.

Automation and Efficiency through Smart Contracts: Smart contracts, powered by blockchain, can automate various asset management tasks, such as initiating maintenance procedures based on certain conditions or transferring ownership of assets. AI can add a layer of intelligence to these smart contracts, enabling them to make complex decisions based on various factors and execute tasks more efficiently.

Enhanced Security and Trust: Blockchain's decentralization and cryptographic security features, combined with AI's ability to detect anomalies and potential threats, can significantly enhance the security of asset management systems. This can foster greater stakeholder trust, a crucial factor in successful asset management.

Creation of Decentralized Autonomous Organizations (DAOs): The combination of AI and blockchain can lead to the creation of DAOs in asset management. These are self-governing, largely automated organizations that operate based on pre-set rules encoded as smart contracts. AI can aid in decision-making processes within these organizations, creating a highly efficient, transparent, and secure system for managing assets.

While the convergence of AI and blockchain holds immense potential, it also presents challenges, such as technical complexity, interoperability issues, and regulatory considerations.

Despite these challenges, the symbiosis of AI and blockchain is poised to revolutionize asset maintenance and management, driving innovation and value creation in the process.

As we look toward the future of asset maintenance and management, it's clear that the intersection of artificial intelligence (AI) and blockchain technology is poised to drive substantial change. The integration of these two transformative technologies will likely usher in a new era characterized by increased efficiency, transparency, predictability, and security.

In terms of efficiency, combining AI's data processing capabilities and blockchain's immutable record-keeping can streamline the asset management process, reducing costs and improving decision-making. This convergence may lead to unprecedented automation in asset management, with AI algorithms identifying maintenance needs and blockchain-based smart contracts initiating the appropriate actions autonomously.

Transparency and traceability, central to effective asset management, are inherent in the design of blockchain technology. This, paired with AI's capacity for in-depth analysis, can provide a detailed and trustworthy view of asset histories, performance metrics, and more.

Predictive maintenance, enabled by AI, can revolutionize asset management by forecasting equipment failures and scheduling maintenance before problems arise, thereby minimizing downtime and enhancing overall productivity. Blockchain's role in this context would be to ensure that the data utilized for these predictions is accurate, consistent, and secure.

On the security front, blockchain's robust encryption protocols combined with AI's anomaly detection capabilities can offer a robust defense mechanism, protecting valuable asset data from cyber threats.

Looking ahead, we can anticipate the rise of Decentralized Autonomous Organizations (DAOs) in asset management. These self-governing entities, built on blockchain and driven by AI, could manage assets efficiently and transparently, minimizing human intervention and bias.

However, despite the promising potential, challenges loom on the horizon. These include technical complexities, regulatory uncertainties, and the need for standardization and interoperability. It will be crucial for stakeholders to address these issues to unlock AI and blockchain's full potential in asset maintenance and management.

In conclusion, the future of asset maintenance and management is undeniably exciting. The symbiosis of AI and blockchain is set to catalyze a paradigm shift in managing and maintaining assets. As these technologies continue to evolve and mature, we can look forward to a future of asset management that is smarter, more secure, and more efficient than ever.

Case Studies and Practical Strategies

Asset Management Challenges and Solutions:

In this chapter, we delve into the main challenges faced in asset management and how modern solutions address them, particularly those bolstered by artificial intelligence.

Challenge 1: Asset Lifecycles and Depreciation

One of the principal challenges in asset management is tracking the lifecycle of an asset and accounting for its depreciation. Assets have a finite lifespan, and their value decreases over time due to wear and tear, obsolescence, and other factors.

Solution: Modern asset management systems can automate tracking asset lifecycles and calculating depreciation. AI algorithms can predict the remaining useful life of assets based on historical data and current condition information, thereby aiding in proactive maintenance and replacement planning.

Challenge 2: Maintenance Management

Ensuring the timely maintenance of assets is another significant challenge. Delayed or missed maintenance can result in asset failure, production downtime, and increased costs.

Solution: AI and IoT technologies enable predictive maintenance, where potential problems are identified before they lead to failure. Sensors can monitor asset conditions in real-time, and AI algorithms can analyze this data to predict when maintenance will be required.

Challenge 3: Asset Visibility and Tracking

Asset visibility refers to knowing where assets are at all times, how they are being used, and by whom. This is crucial in large organizations where assets are spread across multiple locations.

Solution: Asset management systems can track assets using barcodes, RFID tags, or GPS trackers. IoT devices can provide real-time data on asset location and usage. At the same time, AI can analyze this data to identify trends, anomalies, and opportunities for optimization.

Challenge 4: Compliance with Regulations and Standards

Organizations must comply with various regulations and standards related to asset management, such as safety standards, environmental regulations, and accounting rules.

Compliance can be complex and time-consuming, especially for organizations with a large number of assets or those operating in highly regulated industries.

Solution: Modern asset management systems can help organizations maintain compliance by automating record-keeping, providing alerts for non-compliance issues, and generating reports for audits and inspections. AI can further enhance compliance by predicting potential compliance issues based on historical data and current trends.

References

Physical Asset Management Handbook" by John S. Mitchell

https://www.amazon.com/Physical-Asset-Management-Handbook-Mitchell/dp/098536193X

Institute of Asset Management (IAM) - https://theiam.org/

Sustainable Asset Management Linking Assets, People, and Processes for Results by

Roopchan Lutchman, PEng, MBA, North American Leader – Strategic Business Consulting, GHD

https://www.destechpub.com/product/sustainable-asset-management/

Towards a Climate-Neutral Europe

https://climate.ec.europa.eu/system/files/2020-01/toward_climate_neutral_europe_en.pdf

A Whirlwind Tour of Global Carbon Markets, with Stefano De Clara

https://www.resources.org/resources-radio/a-whirlwind-tour-of-global-carbon-markets-with-stefano-de-clara/

The EU Emissions Trading System

https://www.sallan.org/pdf-docs/ets_handbook_en.pdf

ESG Investing and Analysis

https://www.cfainstitute.org/en/research/esg-investing

Hyperledger Foundation

https://www.hyperledger.org/

Predictive Maintenance of Pumps Using Condition Monitoring by Raymond S Beebe

https://shop.elsevier.com/books/predictive-maintenance-of-pumps-using-condition-monitoring/beebe/978-1-85617-408-4

IoT Predictive Maintenance and Optimization

https://www.ibm.com/docs/en/SS7TH3_1.0.1/com.ibm.pmo.doc/pdf/m_pmo_master.pdf

Asset Management Regulations and Standards (ISO 55000)

https://www.iso.org/standard/55088.html

Asset Management: A Best Practices Guide

https://efcnetwork.org/wp-content/uploads/2013/08/Asset-Management-A-Best-Practices-Guide.pdf

Asset Management for Local Officials

https://efcnetwork.org/wp-content/uploads/2013/07/Asset-Management-for-Local-Officials.pdf

Asset Management Tools

https://www.epa.gov/sites/default/files/2020-06/documents/reference_guide_for_asset_management_tools_2020.pdf

Incorporating Asset Management Planning Provisions into NPDES Permits

https://www.epa.gov/sites/default/files/2018-01/documents/incorporating-asset-mgmnt-through-permits.pdf

Checkup program

https://www.epa.gov/sites/default/files/2015-10/documents/assetmgt101.pdf

https://www.epa.gov/sites/default/files/2015-10/documents/cupssusersguide.pdf

State Asset Management Initiative

https://www.epa.gov/sites/default/files/2019-03/documents/asset_management_initiatives_document_508.pdf

FEDERAL REAL PROPERTY ASSET MANAGEMENT

https://www.gao.gov/assets/gao-19-57.pdf

https://www.american.edu/policies/information-technology/upload/it-asset-management-policy-web-2.pdf

Municipality Asset Management Policy

http://www.drakenstein.gov.za/docs/Documents/policy%2008.%20Asset%20Management%20Policy%20Amended%202018.pdf

Australia

https://www.dpie.nsw.gov.au/__data/assets/pdf_file/0005/469526/Asset-Management-Policy.pdf

United Nations Population Fund – Asset Management Policy

https://www.unfpa.org/sites/default/files/admin-resource/FASP_Fixed%20Asset%20Management.pdf

FACT SHEET Asset Management for Sewer Collection Systems

https://www3.epa.gov/npdes/pubs/assetmanagement.pdf

STRATEGIC ASSET MANAGEMENT PLAN 2022

https://www.bpa.gov/-/media/Aep/finance/asset-management/management-plans/2022-transmission-samp.pdf